Land Subsidence Induced by the Engineering-Environmental Effect

Zhen-Dong Cui

Land Subsidence Induced by the Engineering-Environmental Effect

Springer

Zhen-Dong Cui
State Key Laboratory for Geomechanics
and Deep Underground Engineering
School of Mechanics and Civil
Engineering
China University of Mining and Technology
Xuzhou, Jiangsu
China

Funded by National key research and development program (Grant No. 2017YFC1500702), National Natural Science Foundation of China (Grant No. 51208503) and Outstanding Innovation Team Project in China University of Mining and Technology (Grant No. 2015QN001)

ISBN 978-981-13-5693-3 ISBN 978-981-10-8040-1 (eBook)
https://doi.org/10.1007/978-981-10-8040-1

Softcover re-print of the Hardcover 1st edition 2018

Printed on acid-free paper

This Springer imprint is published by the registered company Springer Nature Singapore Pte Ltd. part of Springer Nature
The registered company address is: 152 Beach Road, #21-01/04 Gateway East, Singapore 189721, Singapore

Preface

Land subsidence exists in the cities throughout the world. With the increase of the high-rise buildings, land subsidence induced by the human factors accelerates. Though land subsidence occurs slowly, the accumulation results in the crack of the road, the damage of the pipelines, the tilt or the crack of the buildings, and so on. It causes the loss of the economy. With a large number of high-rise buildings being built and the dewatering of foundation pit for construction of underground structures, the engineering-environmental effect comes to be the main cause of land subsidence. In this monograph, the engineering-environmental effect refers to loads of the high-rise building and the dewatering of foundation pit for the construction of underground structures.

Supported by National key research and development program (Grant No. 2017YFC1500702), National Natural Science Foundation of China (Grant No. 51208503) and Outstanding Innovation Team Project in China University of Mining and Technology (Grant No. 2015QN001), this monograph studies the land subsidence caused by the engineering-environmental effect. With a large number of high-rise buildings being built and the dewatering of foundation pit for construction of underground structures, the engineering-environmental effect comes to be the main cause of land subsidence. The mechanism of land subsidence is studied by the theoretical analysis, the in-site monitoring, the general model test, and the centrifugal model test combined with the scanning electron microscope test (SEM) and the mercury intrusion porosimetry test (MIP).

Chapter 1 is introduction. First, it introduces the progresses of land subsidence in China. Then, the recent researches about land subsidence from four aspects, including in-site monitoring of land subsidence, model tests of land subsidence, theoretical analysis of land subsidence, and prediction of land subsidence, are summarized. Hereby, it points out the purpose and research strategy in this monograph.

Chapter 2 is theoretical analysis of land subsidence caused by the engineering-environmental effect. Land subsidence caused by decompression of confined water is comprised of deformation of aquifer, aquitard, and phreatic stratum. When the decompression features short duration, the consolidation deformation of aquitard is little and the aquitard is supposed to be watertight. Based on displacement coordination condition, theory for elastic-layered system is adopted to calculate layered soil deformation caused by decompression of confined aquifer, including vertical and horizontal components. The formulas for calculating the land subsidence caused by single pile are deduced for different lateral frictions of pile.

Chapter 3 is consolidation of saturated multilayered soils caused by pumping from the dewatering well. Axisymmetric consolidation of multilayered soils induced by groundwater extraction from a dewatering well is studied with consideration of the anisotropic permeability and the well length. Laplace–Hankel transforms are utilized to solve the governing equations. The analytical layer-element method is used to build relationships between displacements, stresses, excess pore pressure, and seepage velocity in the transformed domain. The real solutions can be obtained by the inversion of Laplace–Hankel transforms. A series of parametric studies, especially the length of a dewatering well and the combined effect of pumping and recharging, are conducted to analyze the consolidation behaviors of layered soils.

Chapter 4 is in-site monitoring of land subsidence. The land subsidence caused by four high-rise buildings including World Financial Center (SWFC), Bank of China Tower (BOC Tower), Shanghai Merchants Tower, and Jinmao Tower, which are located in Lujiazui area of Shanghai, is studied by the in-site monitoring data. In addition, the relationship between the changing of the groundwater table and the deformation characteristics of soils is analyzed. The numerical simulation is conducted to study the land subsidence caused by two high-rise buildings with relatively close distance, Jinmao Tower and World Financial Center.

Chapter 5 is centrifuge modeling of land subsidence caused by high-rise building group. This chapter discusses the engineering-environmental effect of high-rise building group on land subsidence under the typical geological subsurface of Shanghai by the centrifuge model tests. The problems studied include the deformation characteristics of different soil layers, the influence of high-rise building group on land subsidence of its central and circumjacent areas, the land subsidence affected by the different building distances, and the stress variation of soil layers due to the engineering-environmental effect of the high-rise building group.

Chapter 6 is microstructure of different soil layers before and after centrifuge modeling of land subsidence caused by the high-rise building group. This chapter studies the microstructure of each soil layer under the building loads in the centrifuge model by SEM and MIP, for qualitative analysis and quantitative analysis, respectively. Before and after the centrifuge model test conducted in Chap. 5, samples of each soil layer are prepared for studying the microstructure of soils by SEM test and MIP test.

Chapter 7 is microstructures of the soil layer at different depths in the centrifuge modeling of land subsidence caused by the interaction of two high-rise buildings. In this chapter, oven-drying method is used to dehydrate the clay samples due to test condition restriction. Many factors at microlevel influencing consolidation of clay are studied, and the MIP is used to examine microstructural change of kaolin clay at different depths. The Menger fractal dimension model and the thermodynamics fractal dimension model are used to analyze MIP data.

Chapter 8 is physical model test of layered soil subsidence considering dual effects of building load and groundwater withdrawal. In this chapter, the deformation of each soil layer caused by high-rise buildings and groundwater exploitation simultaneously is studied. The physical model test based on the typical geological background in Shanghai is conducted, and the particle image velocity (PIV) is adopted to measure the displacement of each point in soil layers.

Chapter 9 is floor area ratio ANFIS model affected by causes of land subsidence. This chapter has demonstrated that the geological environmental capacity of the building is mainly controlled by the land subsidence and the relationship can be assessed using the floor area ratio. The adaptive neuro-fuzzy inference system (ANFIS) is used to evaluate the floor area ratios of four typical areas in Shanghai in order to offer some guidance in respect of urban planning.

Chapter 10 is conclusions and prospects. This part comprehensively summarizes the research conclusions. Several controversial issues are discussed, and then the further research work and prospects are simply described.

This monograph has been prepared with the combined effort of all researchers in the group under Prof. Zhen-Dong Cui's leading, in which Ya-Jie Jia, Zheng Li, Jia-Qiang Yang, Ling-Zi Zhao, Zhong-Liang Zhang, Peng-Peng He, Qiang Yuan, Jun Tan, Cheng-Lin Zhang, Shi-Xi Ren, and some other students all have involved in this comprehensive research work in this monograph. In addition, I should acknowledge the guidance of Prof. Yi-Qun Tang in Tongji University, the guidance of Prof. C. W. W. Ng in Hong Kong University of Science and Technology, and the guidance of Prof. Ronald Y. S. Pak in University of Colorado Boulder.

Xuzhou, China Prof. Dr. Zhen-Dong Cui
January 2018

Contents

About the Author

Prof. Dr. Zhen-Dong Cui is the Dean of Institute of Geotechnical Engineering, China University of Mining and Technology. In July 2008, he obtained the Ph.D. from School of Civil Engineering, Tongji University, Shanghai, China. Since then, he had been a postdoctoral research fellow in the Hong Kong University of Science and Technology for 1 year. He joined Shanghai Institute of Geological Survey in 2009. In July 2010, he joined CUMT as an associate professor and was promoted to full professor in 2013. Supported by China Scholarship Council, from August 2015 to August 2016, as a visiting scholar, he researched and studied in the Department of Civil, Environmental and Architectural Engineering, University of Colorado Boulder.

He won Nomination of 100 Excellent Doctoral Dissertations in China in 2011. He won Shanghai Excellent Doctoral Dissertations in 2010 and Excellent doctorate thesis of Tongji University in 2009. In 2015, he was awarded the third prize of Shanghai Natural Science. In 2013, he was awarded the second prize of Natural Science of the Ministry of Education. In 2008, he was awarded the second prize of Progress of Science and Technology in Shanghai. In 2012, he was selected as Qinglan Project for Outstanding Young Teachers of Jiangsu Province and in 2016, he was selected as 333 Talent Project in Jiangsu Province. In 2014 and in 2017, he was twice selected as Young

Academic Leader of China University of Mining and Technology. In 2015, he was awarded as Excellent Innovation Team Leader of China University of Mining and Technology.

His research interests focus mainly on the urban subway tunnel, the deformation of soft foundation, the soil dynamics, the centrifuge model test, and the artificial frozen soil. In the past 5 years, he took charge of National Natural Science Foundation of China (NSFC), Jiangsu Natural Science Foundation of China, Outstanding Innovation Team Project in China University of Mining and Technology, and Special Fund for China Postdoctoral Science Foundation. He published more than 60 papers, in which 37 English papers (indexed by SCI) have been published in Engineering Geology, Natural Hazards, Computers and Geotechnics, Cold Regions Science and Technology, Environmental Earth Sciences, International Journal of Rock Mechanics and Mining Sciences, etc. He has applied for six national invented patents, among which four patents have been awarded. He is the specialized committee member of the Soft Soil Engineering of the Geotechnical Engineering branch of the Civil Engineering Society of China, the member of International Society for Soil Mechanics and Geotechnical Engineering (ISSMGE) and the member of International Association for Engineering Geology and Environment (IAEG).

List of Figures

List of Tables

Chapter 1
Introduction

1.1 Background

Land subsidence is an environmental geology phenomenon that the land surface elevation falls down due to ground compression (Cui 2012; Hu et al. 2004). Land subsidence in the soft soil area has characteristics of the slow generation, the long duration, the wide ranges of affection, the complex mechanisms of generation, and the difficult prevention (Cui 2008). It is a slow movement and it may break the pipelines and subsidize and/or tilt the buildings in the affected area. The economic loss induced by land subsidence in Shanghai has reached over 46 billion dollars, which means that each millimeter subsidence costs Shanghai nearly 25 million dollars. More than 60 countries are suffering from the costly land subsidence. The cities experienced serious land subsidence, such as Shanghai, Tianjin in China; Osaka, Niigata, and Tokyo in Japan; California region in America; Venice in Italy, and so on.

1.1.1 Land Subsidence of Shanghai

There are two first cities in China to suffer serious subsidence, Shanghai in 1921 and Tianjin in 1923. The land subsidence of two cities shows different characteristics.

Figure 1.1 illustrates the variations of land subsidence with time and Fig. 1.2 shows the comparison between the measured and the calculative subsidence of bedrock mark No. 0–264 in Shanghai (Shen and Xu 2011). At the early stage in Shanghai, the main cause of land subsidence was the irrational withdrawal of groundwater. The extraction of groundwater was mainly conducted in the second and the third aquifers, accounting for 86% of total exploitation quantity. By 1965, the average subsidence in urban area was 1.76 m and the maximum subsidence was 2.63 m. Then, the aquifer for extraction of groundwater was adjusted to the fourth and the fifth aquifers, accounting for 85% of total exploitation quantity and the

Z.-D. Cui, *Land Subsidence Induced by the Engineering-Environmental Effect*,
https://doi.org/10.1007/978-981-10-8040-1_1

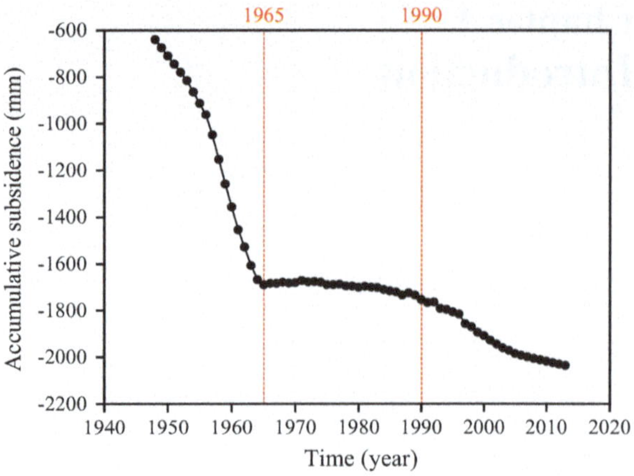

Fig. 1.1 Variations of accumulative land subsidence with time in Shanghai

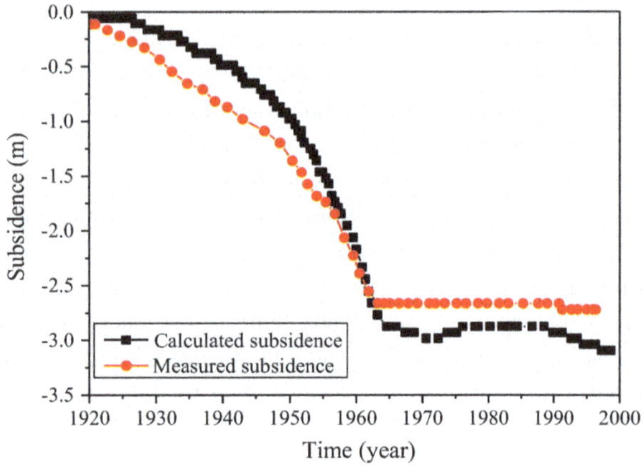

Fig. 1.2 The measured and calculated subsidence of bedrock mark No. 0–264

water was recharged into the second and the third aquifers, which caused the rate of subsidence to be gentle. In the middle of 1970s, the rebound phenomenon of land subsidence appeared. But from 1990s, the rate of land subsidence accelerated gradually.

What are the causes of this phenomenon? The groundwater was exploited in the deeper aquifer and especially from 1990s the quantity of exploitation increased. The recharge of groundwater in the urban area was not enough because the groundwater was exploited not in the urban area but in the suburb and the delayed drainage in the watertight layer is the main cause for the aggravation of land subsidence. However, from some researches, the large-scale engineering construction is the main cause of land subsidence, including the building loads, the pit excavation, the pit dewatering, the shield tunneling, and the leakage of tunnel. Figure 1.3 illustrates the relationship between the accumulative subsidence and the building area in Shanghai (Yang and Gong 2010).

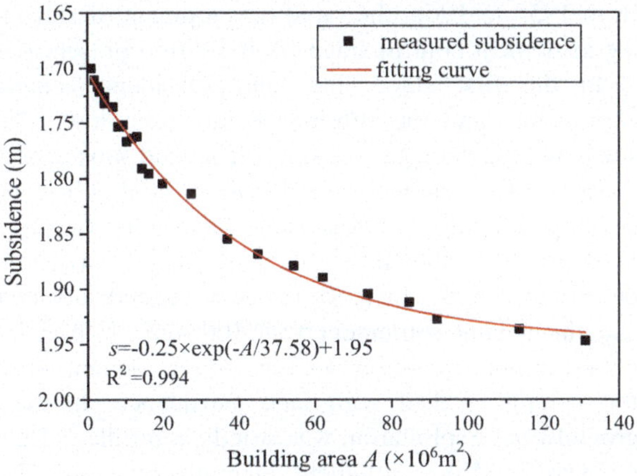

Fig. 1.3 Variations of the accumulative subsidence with the building area in Shanghai

The deformation of soil layers with the depth 75 m from the ground was mainly caused by the engineering construction and the withdrawal of groundwater mainly resulted in the deformation of soil layers with the depth larger than 75 m from the ground (Shi 1999). From the measured data of land subsidence in Shanghai, the land subsidence caused by the engineering construction accounted for 30% of the total land subsidence, but there was less theoretical accordance. On the one hand, the leakage recharge was ignored and the influence of groundwater exploitation on the shallow compression layer was reduced. On the other hand, the pile foundation was chosen by the high-rise buildings. The pile depth of Jinmao Building finished in 1999 was 83 m and that of World Financial Center finished in 2008 was 78 m, but the buried depth of the second confined aquifer is about 50–70 m. The pile foundation reached the influence area of groundwater exploitation and recharge. So the analysis also underestimated the effect of building loads on the deep compression layer. The influence between the groundwater exploitation and high-rise building loads crosses each other (Jia 2015).

In addition, some scholars (Wang and Miao 2011) pointed out that the land subsidence increased slightly, showing typical creep characteristics and the creep of soil is also the cause of long-term development of land subsidence in Shanghai.

The comprehensive effects of various factors lead to the slow growth of land subsidence in Shanghai. However, there is no exact theoretical basis for evaluating the proportion of the land subsidence caused by each factor to the total land subsidence.

1.1.2 Land Subsidence of Tianjin

The development of land subsidence in Tianjin was divided into four stages: the initial subsidence stage from 1923 to 1957, the initial formation of settlement center from 1958 to 1966, the rapid development of subsidence from 1966 to 1985, and the settlement control stage from 1986 to present (Dong et al. 2007).

In the first stage, the land subsidence coincided with the groundwater exploitation and the subsidence rate was slow. The cumulative subsidence was small and the damage was small. The land subsidence was not recognized. With the increase of groundwater exploitation, land subsidence was found in 1958 and it developed rapidly. Some subsidence centers were formed, such as Hebei street and Dazhigu. With the further increase of groundwater exploitation, land subsidence developed rapidly. In 1985, the accumulated subsidence in Tanggu area was 3.1 m and the annual settlement was 100 mm. Tianjin is the coastal city owning the lowest ground elevation in China and its elevation is only 3–5 m. It is more serious for Tianjin to deal with land subsidence. In the land subsidence center, the groundwater exploitation was strictly controlled. The pore water pressure gradually recovered and the settlement phenomenon eased. However, the groundwater level continued to decline and the land subsidence continued to develop in the coastal area. Figure 1.4 illustrates variations of the pore water pressure with the soil layer depth in the urban area and coastal area of Tianjin, respectively (Wu et al. 1998).

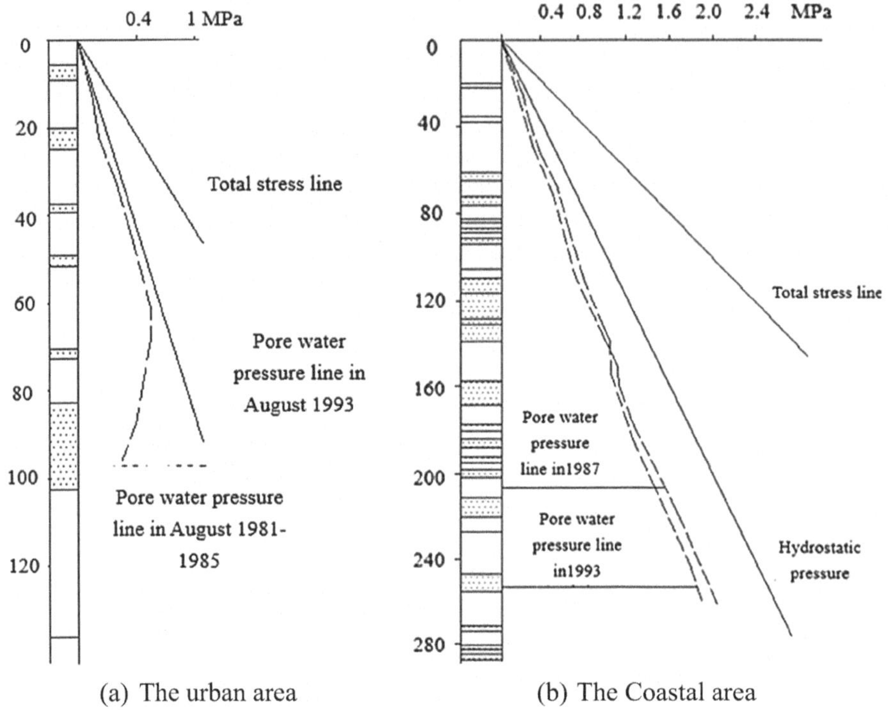

Fig. 1.4 Variations of pore water pressure with depth in Tianjin

Among the factors causing the land subsidence of Tianjin, overexploitation of groundwater is the main factor. The other factors include the natural consolidation of the under consolidation soil whose contribution to the land subsidence of Tianjin is 10–20 mm/a, the exploitation of gas and petroleum whose contribution is 6 mm/a, the engineering construction, the geological tectonic movement, the sea level rise, and so on. At the early stage, the latter three factors affect the land subsidence little. However, at the late stage, when the groundwater exploitation is controlled reasonably, these factors may become main factors resulting in the land subsidence (Xu et al. 2005).

1.1.3 Land Subsidence of Other Areas in China

After the land subsidence appeared in Shanghai and Tianjin, at present there have being 17 provinces experiencing land subsidence in China because of the large-scale industrialization, the urbanization construction, the population agglomeration to the city, the increasing demand for industrial water and domestic water, and the large-scale urban infrastructure. The total area exceeds $5 \times 10^7\ m^2$, where mainly distributes in the Yangtze River Delta Plain, the North China Plain, the Northeast Plain, the Southeast Coastal Plain, the Jianghan Plain, and the Fault Basin, as shown in Fig. 1.5.

Fig. 1.5 Land subsidence in China

The characteristics of land subsidence in each area are summarized in Table 1.1 (Hu et al. 2011; Zhang and Wei 2005; Duan 1998). The large-scale land subsidence was mainly caused by the groundwater exploitation. There was a high correlation between the state of groundwater and the magnitude of land subsidence. At present, the state of groundwater was changed by human activities from two aspects, including the exploitation and utilization of groundwater and the engineering dewatering. Compared with the groundwater exploitation, the range of land subsidence caused by the engineering dewatering was smaller. The drawdown depth of groundwater was related to the excavation scale and the geological condition of the foundation pit. The drawdown depth of groundwater did not exceed 30 m in common. For large foundation pit dewatering project, the recharge, water curtain, and other measures to reduce its influence range were conducted in the same time. After the completion of the project, with the recovery of groundwater level, there was a certain amount of rebound for land subsidence.

In order to reduce the land subsidence caused by groundwater exploitation, the groundwater exploitation was reduced and the recharge was conducted in the areas experiencing larger land subsidence, such as Shanghai, Tianjin, and so on. The groundwater recharge was always larger than the amount of exploitation in the urban area of Shanghai. According to the East satellite TV report, the groundwater exploitation in the urban area of Shanghai was 10^6 m^3 and the amount of recharge was 10^7 m^3 in 2011. Figure 1.6 illustrates the variations of groundwater exploitation recharge with time in Shanghai. The groundwater exploitation reduced in the whole area of Shanghai, from 1.2×10^8 m^3 in 1998 to 1.094×10^7 m^3. However, the amount of recharge increased from 1.8×10^7 m^3 in 1998 to 1.935×10^7 m^3. In 2011, the amount of recharge was 1.861×10^7 m^3, which first exceeded the groundwater exploitation 1.351×10^7 m^3. Although the net exploitation of groundwater was minus, the land subsidence still developed and the mean rate of land subsidence in Shanghai was 6 mm/a.

1.2 Research of Land Subsidence

The research status of land subsidence includes four aspects, including in-site monitoring of land subsidence, model tests of land subsidence, theoretical analysis of land subsidence, and prediction of land subsidence.

1.2.1 In-Site Monitoring of Land Subsidence

Land subsidence monitoring methods include the precise leveling, the global positioning system measurement, the interferometric synthetic aperture radar

Table 1.1 The characteristics of land subsidence in each area

The area experiencing land subsidence	Provinces and cities	Characteristics	Causes
Yangtze river delta region	Shanghai City	Land subsidence first appeared in 1921 and developed quickly during 1950s and 1960s. The maximum cumulative subsidence reached 1.63 m in 1963. Then, the groundwater exploitation was adjusted, and the land subsidence was controlled well. Since 1990, the land subsidence has been developing rapidly again	Groundwater exploitation and engineering construction
	Su-Xi-Chang area in Jiangsu	The land subsidence was found in Su-Xi-Chang area in 1960s. The maximum rates of land subsidence are 40–50, 15–25, and 40–50 mm/a, respectively. The cumulative land subsidence is 1.45, 1.14, and 1.10 m, respectively	Groundwater exploitation
	Hang-Jia-Hu area in Zhejiang	In early 1960s, land subsidence first appeared in Jiaxing. The maximum cumulative subsidence was 0.81 m, and the maximum rate of subsidence was 41.9 mm/a. In recent years, the rate is about 28 mm/a	Groundwater exploitation
North China plain	Tianjin City	The urban area, Tanggu, and Hangu were subsidence centers, and the maximum cumulative subsidence was 3.92 m. The subsidence rate of the urban area was 70–110 mm/a in 1985	Exploitation of groundwater, gas, and petroleum; geological tectonic movement; natural consolidation of soils
	Beijing City	The maximum land subsidence was 0.85 m, and the ground fissures were formed	Groundwater exploitation
	Hebei Province	The land subsidence first appeared in Hebei Plain in 1950s. At present, there are 10 subsidence centers, including Cangzhou, Hengshui, Renqiu, Hejian, Bazhou, Baoding, Dacheng, Nangong, Feixiang, and Handan Cangzhou is the biggest subsidence center and its maximum cumulative subsidence is 1.13 m	Groundwater exploitation
	Henan Province	The maximum subsidence of Kaifeng, Luoyang, Xuchang, and Anyang was 0.21, 0.11, 0.21, and 0.34 m, respectively. The maximum rate of subsidence of Anyang was 65 mm/a	Groundwater exploitation
	Anhui Province	The maximum drawdown of water level was 60–70 m in Fuyang, and the maximum subsidence was 1.2 m	Exploitation of groundwater and solid mineral resources
	Shandong Province	The cumulative land subsidence of Dezhou, Jining, and Heze was 0.3, 0.2, and 0.07 m, respectively, and their subsidence rates were 32, 25.2, and 9.68 mm/a, respectively	Groundwater exploitation

(continued)

Table 1.1 (continued)

The area experiencing land subsidence	Provinces and cities	Characteristics	Causes
Northeast plain	Three provinces in Northeast China	There existed land subsidence in Harbin, Qiqihar, Changchun, Shenyang, and Daqing	Exploitation of groundwater and petroleum
Jianghan plain	Hubei Province	There existed seriously land subsidence in Wuhan and Xiaogan	Groundwater exploitation and geological tectonic movement
Fault Basin	Shaanxi Province	The subsidence center was found in Xi'an in 1960s, and the maximum cumulative subsidence was 1.94 m. The maximum and mean rates of subsidence were 300 and 80–126 mm/a, respectively. There were 11 ground fissures	Groundwater exploitation and geological tectonic movement
	Shanxi Province	There existed land subsidence in Datong in early 1980s, and there were eight ground fissures. The maximum land subsidence was 1.97 m in Taiyuan, and its rate was 37–114 mm/a. There also existed serious land subsidence in Yuci and Jiexiu	Groundwater exploitation and geological tectonic movement
Southeast coastal plain	Zhejiang Province	The land subsidence first appeared in Ningbo in 1960s. The maximum cumulative subsidence and the maximum subsidence rate were 0.46 m and 18 mm/a, respectively	Groundwater exploitation
	Fujian Province	Land subsidence was found in Fuzhou in 1957. The maximum cumulative subsidence and the subsidence rate were 0.69 m and 2.9–21.8 mm/a, respectively	Groundwater exploitation
	Guangdong Province	The maximum land subsidence in Zhanjiang was 0.11 m, and the maximum subsidence rate was 4.09 mm/a	Groundwater exploitation
	Taiwan Province	Zhangbei, Yunlin, Jiayi, Tainan, Gaoxiong, and Pingdong experienced the largest land subsidence were 1.53, 2.05, 1.17, 0.63, 0.25, and 3.12 m, respectively	Groundwater exploitation

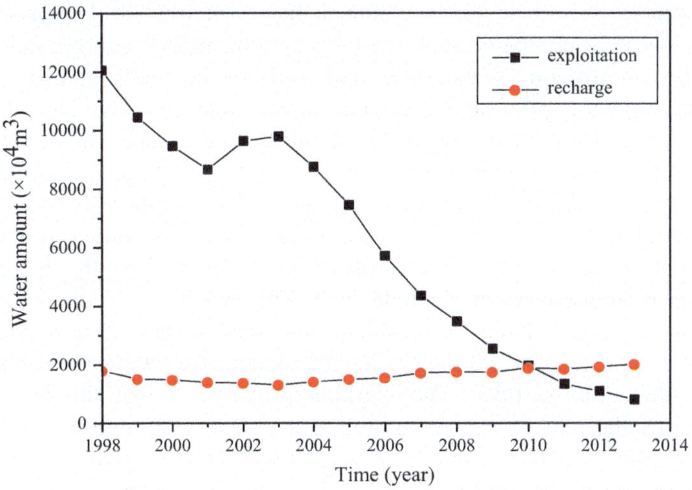

Fig. 1.6 Variations of groundwater exploitation and recharge with time in Shanghai

measurement, the bedrock mark and layered mark measurement, and the laser radar measurement.

Precise leveling requires a reference point first. It is generally considered that the datum point does not produce vertical displacement. Shanghai has built a large number of bedrock marks as leveling control points. Under the premise of ensuring the measurement accuracy, the leveling is conducted to the designated leveling points in the study area. Through the adjustment calculation of the leveling network, the elevation of leveling points can be obtained. The elevation of the leveling point is measured every other period of time, so that the accumulated settlement of the leveling point in a period of time can be obtained. The accuracies of the first-order leveling and the second-order leveling are 1 and 2 mm/km, respectively.

The global positioning system (GPS) measurement needs to build the GPS station on the ground. The GPS station determines the distance between the satellite and the GPS station according to the signal released by the satellite. Then, the three-dimensional coordinates of the station at a certain time are determined according to the location of several satellites. At present, the GPS stations have been built in Tianjin, Shanghai, Taiwan, Su-Xi-Chang, and so on. The resolution reaches 3 mm. In addition, the horizontal displacement of the measuring point can be obtained by using the GPS system.

The interferometric synthetic aperture radar (InSAR) is used to measure the SAR image of the same region at different times. The phase difference of the two images is calculated to obtain the interference image. The ground deformation contour map is obtained by ground deformation measurement. For the permanent scatterer interferometry (PSI), the decorrelation of time and space and the atmospheric effects which limit the accuracy of the measurement are well handled and the accuracy can reach mm level. In China, it has been used in Tianjin, Shanghai, Suzhou, and so on, compared with the GPS system, InSAR can get more data.

The standard structure of bedrock mark and layered mark includes protection device, guide device, pilot device, ground device, and guide device. The base of bedrock mark is fixed in the bedrock, so the bedrock mark can be used as the control point in leveling. However, the base of the layered mark contacts with the interface of soil layers and it can move up and down with the soil layer, reflecting the rise and fall of the measured soil layer. The technology of one well owing many marks is that the sensing devices are distributed at different depths, to achieve the observation of multilayers in a single well. This method is economical and the precision is up to 0.01–1 mm. It is widely used in Shanghai, Tianjin, and Taiwan.

The light detection and ranging (LiDAR) includes two types: the ground laser radar and the airborne lidar. The working principle is that the laser beam is transmitted by the sensor and transmitted to the ground through the air. After the reflection of the ground, the sensor receives the reflected signal and the distance from the laser to the ground surface can be calculated according to the time of transmitting and receiving the signal. The accuracy of airborne lidar is not high, only being 0.3 m. Ground laser radar is usually used to obtain the parameters of moving objects or to draw the shape of the object surface. For measuring the land subsidence, the technical requirements of the operator are high.

The monitoring of land subsidence develops from the artificial to the intelligence and from single point to multiple points. Although the precise leveling has high precision, it can obtain less data points and it is time-consuming and laborious. Because of the limitation of accuracy and measurement time, LiDAR can only be used for rough topographic mapping. GPS and InSAR can obtain more data points, but the accuracy of measurement needs to be further improved. The land subsidence at different depths in the same measuring point, the technology of one well owing many marks, is more suitable. But a variety of monitoring methods are still needed.

1.2.2 Theoretical Analysis of Land Subsidence

The area of ground subsidence caused by groundwater exploitation is often very large. The thickness of the vertical compression layer is smaller than the scope of the plane. There is the difference of orders of magnitude between the horizontal displacement and the vertical displacement, even near the pumping well. But far away from the pumping well, the difference is greater (Li et al. 2013). Generally, the deformation of soil caused by groundwater exploitation is reduced to one-dimensional compression. According to the coupling relation of seepage and stress, the calculation model can be divided into two kinds: sequential coupling and direct coupling.

For each soil layer, the sequential coupling refers to the continuity equation of water, that is, the difference between the quantity of water flowing into and out of the infinitesimal element in unit time is equal to the increase of the amount of water in the infinitesimal element in unit time, and the change of the soil water pressure with time is combined with the corresponding boundary conditions. Generally, the compression of water is not considered. According to the principle of effective stress, the change of water pressure is transformed into the change of effective stress. Then, the strain and the settlement of soil layer can be obtained by the constitutive relation of soil. The constitutive relation of soil can be taken as elastic, elastoplastic, and viscoelastic plastic according to the characteristics of deformation of soil. The characteristics of deformation are determined by the critical water level in the soil, i.e., the pre-consolidation pressure of the soil. The characteristics of deformation of clay had been studied by many researchers, which was considered as elastic–plastic deformation (Hu et al. 2002). It was found that clay creep accounted for a large proportion of deformation, so the creep of clay should be taken into account (Men 1999). The deformation of sandy soil is elastic. From analyzing the data of the indoor test and the in-site, it was found that there also existed the plastic deformation and the creep for sand (Zhang et al. 2009). The viscoelastic–plastic constitutive model of soil is established, which can be transformed into the elastic model, the viscoelastic model, and the elastoplastic model by adjusting parameters of the model. In the calculation of soil deformation, some scholars did not consider the change of the permeability coefficient of soil with the change of stress state and some studies adjusted the permeability coefficient in the process of the deformation of soil. The sequential coupling method was used to study the land subsidence caused by the groundwater exploitation in Shanghai (Ye et al. 2011), Tianjin (Shearer 1998), Suzhou (Chen et al. 2003), and other areas.

The direct coupling theory is based on Biot consolidation theory and the seepage deformation coupling equation is established, including the equilibrium differential equation, the geometric equation, the constitutive equation, and the continuity equation. The four unknowns of the pore pressure and the displacement of soil in three directions are solved directly. The constitutive equation is determined by the deformation characteristics of the soil, which is as the same as those of the soil in the sequential coupling model. The advantage of the direct coupling is that the horizontal displacement of the soil can be calculated, which can be used in the formation of ground fissures. The disadvantage is that the computation is complex and it is usually solved by numerical method. When the solution area is large, the computation is very complex. Therefore, the direct coupling method is generally used for the calculation of small regional land subsidence (Chen and Li 2006). When calculating the large-scale land subsidence, the soil parameters should be calculated according to the relationship between the actual amount of groundwater exploitation and the subsidence.

For land subsidence caused by groundwater exploitation, especially for the confined aquifer, the change of saturation of soil is not taken into account in common and the soil is always considered in the saturated state. Land subsidence induced by dewatering includes the compression of the aquifer and the consolidation deformation of the weak permeable aquifer (Luo et al. 2006). The compression of permeable layer is calculated directly according to the effective stress of soil before and after depressurization. The compressibility of the weak aquifer is calculated as follows. (1) According to the pore pressure at the boundary eventually being adjusted to the same as the aquifer, the consolidation equation of the weak layer is established to calculate its deformation (Luo et al. 2004). (2) The weak permeable aquifer is regarded as the water-resisting layer and the effect of depressurization of confined aquifer on the overlying soil is equivalent to the downward pulling force to calculate the final deformation of the weak permeable layer (Gong and Zhang 2011). (3) Without considering the existence of overlying weak permeable layer, the compression of permeable layer is regarded as ground subsidence (Zhou et al. 2011). As for whether to consider the compression amount of each soil layer, in addition to considering the time factor and the permeability of soil layer, the thickness of soil layer plays an important role.

The calculation model of land subsidence caused by the building load is much simpler. The deformation is generally regarded as the three-dimensional problem, and the computational theory is Biot consolidation theory. The subsidence can change with time of consideration. In addition, the additional stress can be converted into the effective stress without considering the consolidation process of the soil, and the elastic–plastic or elasto-viscoplastic constitutive relation of the soil is used to calculate the final settlement of the area. The building foundation was taken as the pile foundation and the elastoplastic or viscoelastic–plastic constitutive model was adopted to study the distribution law of the land subsidence under different floor area ratios (Song and Tang 2012). Except for the factors such as the floor area ratio, there is no quantitative study on the influence of the basic form on the distribution of the subsidence in the whole area. Some scholars believed that the role of pile foundation was only to transfer the upper load and changing the pile diameter or pile length only changed the influence range of foundation and the force between piles and soils, which had little effect on the distribution of the whole land subsidence area. But there was no quantitative research conclusion.

For the land subsidence induced by both groundwater exploitation and the building loads, without considering the change of soil saturation and only considering the simple geological condition of single-layer soil, the land subsidence caused by the two factors is linearly superposed (Zhao et al. 2008). Considering the complex geological conditions of the multilayered soil, without considering the change of soil saturation, the subsidence caused by interaction of both the groundwater exploitation and the building loads is smaller than the sum of the land subsidence caused by the groundwater exploitation and that caused by the building loads through the numerical calculation (Ding et al. 2011). There was no detailed description of the basic form of the building, and there was only one building. The order of pumping and building was not clear. The effect of interaction needs to be further studied.

1.2.3 Model Tests of Land Subsidence

The boundary condition of laboratory test is easy to control and the boundary condition and stress state of model test can be changed according to the need of research. The land subsidence caused by the groundwater exploitation has the characteristics of the long duration, the slow development, and the wide range of influence. Even in laboratory tests, the selection of similar materials and the simulation of pumping conditions are quite difficult. The centrifugal model test was conducted to study the land subsidence in the small area caused by groundwater exploitation (Sun et al. 2008), which simulated the plane problem.

The general model test and the centrifuge model test were conducted by Tang et al. (2008) and Cui et al. (2010) to study the land subsidence under the typical engineering geology in Shanghai with the groundwater unchanging, considering the building spacing and the floor area ratio. The land subsidence caused by high-rise buildings was affected by the construction sequence of the buildings. Figure 1.7 illustrates the variations of subsidence of model buildings with time. Building A was the first loaded, and the subsidence was greatest, followed by Building B which was the second to be loaded. Buildings C and D were the last to be loaded at the same time and their subsidence was the least. The stress superimposition effect at the central area was the largest. The land subsidence of the points near to the former building was larger than that of those near to the latter building. The central area of the group of buildings experienced the maximum subsidence. The points within one times the width of the foundation from the center of the building had the second largest subsidence. The least subsidence was recorded in the area two times the width of the foundation from the center of the building.

1.2.4 Prediction of Land Subsidence

The prediction of land subsidence can be divided into three types, including the deterministic model, the stochastic statistical model, and the artificial intelligence model.

According to the deformation mechanism of the soil, the calculation model was established by the deterministic model, considering the engineering geological conditions, the soil constitutive relationship, and the amount of groundwater exploitation. By setting the corresponding boundary conditions, or adjusting the amount of groundwater exploitation, or adjusting the floor area ratio (Song and Tang 2012), the variations of land subsidence with time under different groundwater exploitations or different floor area ratios in a certain area were obtained. According to the local control requirements for land subsidence, the corresponding policy of groundwater exploitation was formulated. The method requires more information and complex computation, but the results are reliable. The land

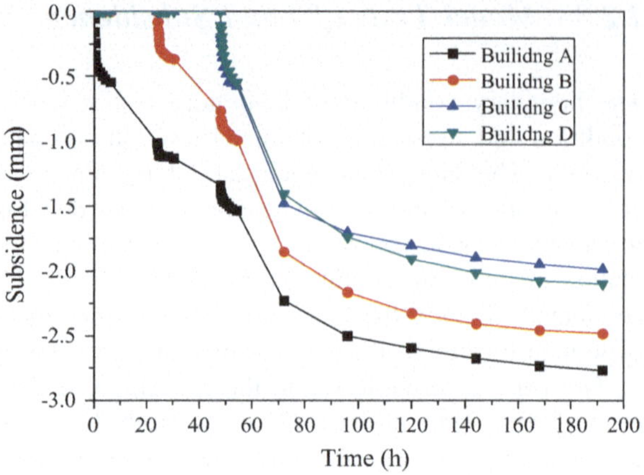

Fig. 1.7 Variations of subsidence of model buildings with time

subsidence in Shanghai, Tianjin, Suzhou, and so on was studied by this method (Ye et al. 2011; Shearer 1998).

The variations of land subsidence with time in a certain area are analyzed by the stochastic statistical model. The mathematical methods include the regression analysis and the gray theory. This method cannot reveal the dynamic mechanism of land subsidence. In addition, the formation of land subsidence is the result of multiple factors, such as the groundwater exploitation, the building load, and other factors. The stochastic statistical model cannot consider the correlation and weight of each factor. This method cannot solve the problem that which factor of the groundwater exploitation and the building load is the main factor to cause land subsidence in Shanghai. However, the model is simple and the conclusion is clear. It can simulate the overall trend of land subsidence and has some guiding significance for policymaking. This method was conducted to predict the land subsidence based on both the in-site monitoring and the mode test by Tang et al. (2008).

The artificial intelligence model includes the artificial neural network, the genetic algorithm, and so on. Compared to the stochastic statistical model, this method can realize the multivariable and nonlinear numerical prediction. This method was used to predict the land subsidence in Tianjin by Li et al. (2005).

1.3 Discussion and Conclusions

Above all research of land subsidence from the in-site monitoring, the theoretical analysis, the model tests, and the prediction of land subsidence, the horizontal subsidence of the soil has been less studied. Compared to the vertical subsidence, the horizontal subsidence of soil caused by either the groundwater exploitation or the building loads is small. However, whether the horizontal subsidence should be neglected or not and its damage should not be considered or not has been less reported.

The cities experienced larger land subsidence such as Shanghai, Tianjin, Tokyo, and so on and decrease the amount of groundwater exploitation and in the meantime increase the amount of groundwater recharge. It is not enough only from the absolute magnitude of the exploitation and the recharge because the levels of the exploiting and the recharging are different.

In addition, there are less studies on the land subsidence from the aspects of the theoretical research, the basic form of buildings, the adjustment of groundwater surface induced by the groundwater exploitation, and the changing of the saturation state of the overlying soil.

The land subsidence includes the compression of different soil layers. However, some studies tend to calculate the deformation of the weak aquifer and some studies consider the compression of the aquifer. The compression of different soil layers should be considered together.

References

Chen P, Li XZ (2006) Application of Biot model in well-point dewatering causing ground movement. Chin J Undergr Sp Eng 2(1):137–140 (in Chinese)

Chen CX, Pei SP, Jiao JJ (2003) Land subsidence caused by groundwater exploitation in Suzhou City, China. Hydrogeol J 11(2):275–287

Cui ZD (2008) Study of the land subsidence caused by the dense high-rise building group in the soft soil area. Ph.D. thesis, Tongji University, Shanghai (in Chinese)

Cui ZD (2012) Land subsidence disaster caused by natural factors and human activities. Disaster Adv 5(2):3–4

Cui ZD, Tang YQ, Yan XX (2010) Centrifuge modeling of land subsidence caused by the high-rise building group in the soft soil area. Environ Earth Sci 59(8):1819–1826

Ding DM, Ma FS, Zhang YM, Wang J, Guo D (2011) Characteristics of land subsidence due to both high-rise building and exploitation of groundwater in urban area. J Eng Geol 19(3):433–439 (in Chinese)

Dong KG, Zhou J, Yu Q (2007) Feature of the land subsidence and its damage in Tianjin City. J Geol Hazards Environ Preserv 18(1):67–70 (in Chinese)

Duan YH (1998) Research status on land subsidence and sustainable development in the 21st century of China. Chin J Geol Hazard Control 9(2):1–5 (in Chinese)

Gong XN, Zhang J (2011) Settlement of overlaying soil caused by decompression of confined water. Chin J Geotech Eng 33(1):145–149 (in Chinese)

Jia YJ (2015) Land subsidence caused by high-rise building group based on the varying groundwater table. Master Degree thesis, China University of Mining and Technology (In Chinese)

Hu RL, Wang SJ, Lee CF (2002) Characteristics and trends of land subsidence in Tanggu, Tianjin, China. Bull Eng Geol Environ 61(3):213–225

Hu RL, Yue ZQ, Wang LC, Wang SJ (2004) Review on current status and challenging issues of land subsidence in China. Eng Geol 76(1–2):65–77

Li T, Pan Y, Lou HJ, Li B, Wang H, Zou LZ (2005) Application of the artificial neural network in land subsidence prediction in the urban area of Tianjin municipality, China. Reg Geol China 24 (7):677–681

Li N, Li M, Zhao FS, Wang QY (2013) Numerical simulation analysis of Xi'an land subsidence by FLAC2D. J Catastrophology 28(3):210–214 (in Chinese)

Luo GY, Pan H, Cao H, Yin XL (2004) Analysis of settlements caused by decompression of confined water. Rock Soil Mech 25(Supp. 2):196–200 (in Chinese)

Luo ZJ, Li L, Yao TQ (2006) Coupling model of three dimensional seepage and land-subsidence for dewatering of deep foundation pit in loose confined aquifers. Chin J Geotech Eng 28 (11):1947–1951 (in Chinese)

Men FL (1999) Preliminary investigations on rheological properties of clay and ground settlement in Shanghai city. J Nat Hazards 8(3):117–126 (in Chinese)

Shearer TR (1998) A numerical model to calculate land subsidence, applied at Hangu in China. Eng Geol 49(2):85–93

Shen SL, Xu YS (2011) Numerical evaluation of land subsidence induced by groundwater pumping in Shanghai. Can Geotech J 48(9):1378–1392

Shi W (1999) Analysis of the effects of engineering construction and groundwater exploitation on land subsidence. Shanghai Geol 72:50–54 (in Chinese)

Song SP, Tang YQ (2012) Best floor area ratio of high density buildings based on rheological theory. J Cent S Univ (Sci Technol) 43(6):2349–2356 (in Chinese)

Sun ZY, Zhang G, Zhang JM, Li GH, Zheng RH (2008) Centrifuge modeling of ground settlement due to groundwater pumping. China Civil Eng J 41(4):67–73

Tang YQ, Cui ZD, Wang JX, Lu C, Yan XX (2008) Model test study of land subsidence caused by high-rise building group in Shanghai. Bull Eng Geol Environ 67(2):173–179

Wang F, Miao LC (2011) Study of long-term deformation characteristics of and aquifers land subsidence caused by groundwater withdrawal. Chin J Rock Mech Eng 30(Supp. 1):3135–3140 (in Chinese)

Wu TJ, Cui XD, Niu XJ (1998) Study of land subsidence in Tianjin municipality and its comprehensive harnessing. Hydrogeol Eng Geol 5:17–20 (in Chinese)

Xue YQ, Zhang Y, Ye SJ (2005) Land subsidence in China. Environ Geol 48(6):713–720

Yang TL, Gong SL (2010) Microscopic analysis of the engineering geological behavior of soft clay in Shanghai, China. Bull Eng Geol Environ 69(4):607–615

Ye S, Xue Y, Wu J (2011) Regional land subsidence model embodying complex deformation. Water Manage 164(10):519–531

Zhang AG, Wei ZX (2005) Land subsidence in China. Shanghai science and Technology Press, Shanghai (in Chinese)

Zhang Y, Xue YQ, Wu JC (2009) Experimental research on creep of Shanghai sands. Rock Soil Mech 30(5):1226–1231 (in Chinese)

Zhao H, Qian H, Li Y, Peng JB (2008) Land subsidence model under dual effects of groundwater pumping and construction loading. J Earth Sci Environ 30(1):57–59 (in Chinese)

Zhou NQ, Tang YQ, Lou RX, Jiang SM (2011) Numerical simulation of deep foundation pit dewatering and land subsidence control of Xujiahui Metro station. Chin J Geotech Eng 33 (12):1950–1956 (in Chinese)

Chapter 2
Theoretical Analysis of Land Subsidence Caused by the Engineering-Environmental Effect

2.1 Introduction

Land subsidence is most often caused by human activities, mainly from the removal of subsurface water. Land subsidence occurs when large amounts of groundwater have been withdrawn from certain types of rocks, such as fine-grained sediments. The rock compacts because the water is partly responsible for holding the ground up. When the water is withdrawn, the rock falls in on itself.

With the withdrawal of groundwater controlled reasonably in many cities throughout the world, the engineering-environmental effect of the dense high-rise building group comes to be the main cause of land subsidence. In the urban region of China, groundwater extraction is the primary cause of land subsidence. But from the 1960s, the withdrawal of groundwater was controlled reasonably and especially from the end of the 1970s, the pumping of groundwater was strictly controlled in the urban area in Shanghai. The quantity of water recharged into the subsurface was always greater than that of the pumping and the extracted aquifers were gradually adjusted. As a result of these measures, the subsidence caused by pumping and recharging kept smooth and gentle in the urban area. During the 1990s, however, with the development of the economy, a variety of municipal works and high-rise buildings were constructed and the subsidence appeared to accelerate in Shanghai (Cui 2012b).

2.2 Land Subsidence Caused by Decompression of Confined Water

Land subsidence caused by decompression of confined water is comprised of compression of aquifer, aquitard, and phreatic stratum (Luo et al. 2006). Methods for calculating land subsidence include sequential coupling model and direct

Z.-D. Cui, *Land Subsidence Induced by the Engineering-Environmental Effect*,
https://doi.org/10.1007/978-981-10-8040-1_2

coupling model (Galloway and Burbey 2011). The sequential coupling model is to calculate soil additional stress according to groundwater flow theory and the principle of effective stress, and then, to calculate soil deformation with certain soil constitutive model. The sequential coupling model was widely used to calculate aquifer compression (Wu et al. 2010; Zhou et al. 2010, 2011; Lou et al. 2011; Shen and Xu 2011; Chen et al. 2013). When decompression of confined aquifer features short duration, the consolidation deformation of aquitard is little. Gong et al. (2011) and Wang et al. (2013) supposed that the adjacent aquitard was watertight and the bottom of upper soil layer should be subjected to the equivalent tension stress of the increasing effective stress value in the aquifer. That is, the stress coordination condition was adopted. And then the Mindlin's solution for displacement was used to calculate the land subsidence. Zhou et al. (2011) considered the ground subsidence to be equal to the aquifer compression. If leakage recharge from adjacent aquitards should be taken into consideration, Terzaghi's one-dimensional consolidation theory can be applied to calculate the adjacent aquitards deformation based on the boundary condition that the adjacent aquitards pore pressure should be equal to the confined aquifer (Luo et al. 2004; Zeng and Wang 2012; Wu et al. 2014). Land subsidence equals to the superposition of the compression of the aquifer, the aquitard, and the phreatic stratum (Miu et al. 1991). The theoretical basis of the direct coupling model is Biot's consolidation theory (Luo et al. 2008, 2011, 2013; Zheng et al. 2014).

When using the stress coordination condition and the Mindlin's solution, it is ambiguous to choose soil physical property indexes, that is, it is uncertain to use the elastic parameters of the aquifer or the upper soil layers. And for the half-space hypothesis, the calculated results tend to be larger. Taking no consideration of the existence of the upper aquitard and phreatic stratum, the influence of upper soil layers on ground deformation cannot be studied so that the obtained land subsidence may not be accurate. The same problem exists in adding the compression of the aquifer, the aquitard, and the phreatic stratum directly to obtain the land subsidence. The solution of direct coupling model is complicated, and the adoption of the numerical method is necessary. What is more, if the study area is too big, numerical method may not help for the complicated calculation. Besides, soil horizontal displacement was rarely studied in the above research.

When decompression of confined aquifer features short duration, consolidation of confined aquifer can complete within a short time for its high permeability. For low permeability, consolidation deformation of the adjacent aquitards within a short time is little. Taking no consideration of consolidation deformation of upper soil layers, displacement of the adjacent aquitards should be equal to confined aquifer. This chapter used theory for elastic-layered system to calculate layered soil deformation based on displacement coordination condition, including vertical and horizontal components, caused by decompression of confined aquifer. The vertical and horizontal displacements of aquifer and overlying soil were studied, and the results were verified by in situ pumping test and numerical simulation results (Cui et al. 2016).

2.2.1 Theory for Elastic-Layered System

As Fig. 2.1 shows, the *Love displacement equation* of single-layer soil subjected to axisymmetric load should satisfy the following compatible equation (Burmister 1944, 1945a, b; Zhang et al. 1985, 1986):

$$\left(\frac{\partial^2}{\partial r^2} + \frac{1}{r}\frac{\partial}{\partial r} + \frac{\partial^2}{\partial z^2}\right)^2 \zeta = 0 \tag{2.1}$$

The displacement and stress components can be described as

$$\left.\begin{array}{l} u_r = -\frac{1}{2G}\frac{\partial^2 \zeta}{\partial r \partial z} \\ w = \frac{1}{2G}[2(1-\mu)\nabla^2 - \frac{\partial^2}{\partial z^2}]\zeta \\ \sigma_z = \frac{\partial}{\partial z}[(2-\mu)\nabla^2 - \frac{\partial^2}{\partial z^2}]\zeta \\ \tau_{rz} = \frac{\partial}{\partial r}[(1-\mu)\nabla^2 - \frac{\partial^2}{\partial z^2}]\zeta \end{array}\right\} \tag{2.2}$$

where u_r is the horizontal displacement, w being the vertical displacement, σ_z being the vertical stress, τ_{rz} being the shear stress, G being the shear modulus, and μ being the Poisson's ratio.

Perform Hankel transformation for Eq. (2.1) and the following ordinary differential equation about z can be obtained,

$$\left(\frac{\mathrm{d}^2}{\mathrm{d}z^2} + \xi^2\right)^2 \overline{\zeta}(\xi, z) = 0, \tag{2.3}$$

where $\overline{\zeta}(\xi, z) = \int_0^\infty \zeta(r, z)\, rJ_0(\xi r)\mathrm{d}r$.

The general solution is

$$\overline{\zeta}(\xi, z) = (A + Bz)\mathrm{e}^{-\xi z} + (C + Dz)\mathrm{e}^{\xi z} \tag{2.4}$$

where A, B, C, and D are constants determined by the boundary condition.

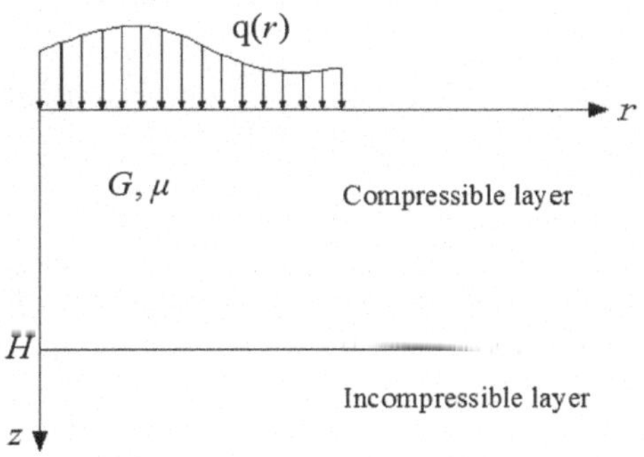

Fig. 2.1 Single-layer soil subjected to axisymmetric load

Perform Hankel transformation for Eq. (2.2) and express stress and displacement distribution function with Eq. (2.4),

$$\left.\begin{aligned}
\overline{u}_r(\xi, z) &= \tfrac{1+\mu}{E}\left[\xi e^{-\xi z}A - (2-2\mu-\xi z)e^{-\xi z}B + \xi e^{\xi z}C + (2-2\mu+\xi z)e^{\xi z}D\right]\\
\overline{w}(\xi, z) &= \tfrac{1+\mu}{E}\left[\xi e^{-\xi z}A + (1-2\mu+\xi z)e^{-\xi z}B - \xi e^{\xi z}C + (1-2\mu-\xi z)e^{\xi z}D\right]\\
\overline{\sigma}_z(\xi, z) &= -\xi^2\left(e^{-\xi z}A + ze^{-\xi z}B + e^{\xi z}C + ze^{\xi z}D\right)\\
\overline{\tau}_{rz}(\xi, z) &= \xi\left[-\xi e^{-\xi z}A + (1-\xi z)e^{-\xi z}B + \xi zC + (1+\xi z)e^{\xi z}D\right]
\end{aligned}\right\} \tag{2.5}$$

where E is the elastic modulus and H is the thickness of soil layers.

Suppose that $z = 0$ in Eq. (2.5), linear algebraic equations about A, B, C, and D can be obtained. Solve the linear algebraic equations and substitute A, B, C, and D in Eq. (2.5); the initial function solution for single elastic soil layer is as follows:

$$\begin{bmatrix} \overline{u}_r(\xi,z) \\ \overline{w}(\xi,z) \\ \frac{\overline{\sigma}_z(\xi,z)H}{E} \\ \frac{\overline{\tau}_{rz}(\xi,z)H}{E} \end{bmatrix} = \begin{bmatrix} F_1 & \frac{1}{4(1-\mu)}(F_5-F_4) & \frac{z(1+\mu)}{4H(1-\mu)}(F_3+F_4) & \frac{1+\mu}{2\xi H(1-\mu)}F_5 \\ \frac{1}{4(1-\mu)}(F_6-F_3) & F_2 & \frac{1+\mu}{2\xi H(1-\mu)}F_6 & -\frac{z(1+\mu)}{4H(1-\mu)}(F_3+F_4) \\ \frac{\xi^2 zH}{4(1-\mu^2)}(F_3+F_4) & \frac{\xi H}{2(1-\mu^2)}F_4 & F_2 & -\frac{1}{4(1-\mu)}(F_5-F_4) \\ \frac{\xi H}{2(1-\mu^2)}F_3 & \frac{\xi^2 zH}{4(1-\mu^2)}(F_3+F_4) & -\frac{1}{4(1-\mu)}(F_6-F_3) & F_1 \end{bmatrix} \times \begin{bmatrix} \overline{u}_r(\xi,0) \\ \overline{w}(\xi,0) \\ \frac{\overline{\sigma}_z(\xi,0)H}{E} \\ \frac{\overline{\tau}_{rz}(\xi,0)H}{E} \end{bmatrix} \tag{2.6}$$

$$\left.\begin{aligned}
F_1 &= ch\xi z + \tfrac{\xi z sh\xi z}{2(1-\mu)}\\
F_2 &= ch\xi z - \tfrac{\xi z sh\xi z}{2(1-\mu)}\\
F_3 &= sh\xi z + \xi z ch\xi z\\
F_4 &= sh\xi z - \xi z ch\xi z\\
F_5 &= (3-4\mu)sh\xi z + \xi z ch\xi z\\
F_6 &= (3-4\mu)sh\xi z - \xi z ch\xi z
\end{aligned}\right\} \tag{2.7}$$

From Eq. (2.6), it is clear that if the four functions, $\overline{u_r}(\xi, 0)$, $\overline{w}(\xi, 0)$, $\overline{\sigma_z}(\xi, 0)$, and $\overline{\tau_{rz}}(\xi, 0)$ are known, the displacement and stress of every point in soil foundation can be determined. Suppose that $\xi H = \alpha$, from Eq. (2.6), the relationship of the top surface displacement and bottom surface displacement can be obtained.

$$\left.\begin{array}{l}\overline{u_r}(\xi,H)=a_{11}\overline{u_r}(\xi,0)+a_{12}\overline{w}(\xi,0)+a_{13}\frac{\overline{\sigma_z}(\xi,0)H}{E}+a_{14}\frac{\overline{\tau_{rz}}(\xi,0)H}{E}\\ \overline{w}(\xi,H)=a_{21}\overline{u_r}(\xi,0)+a_{22}\overline{w}(\xi,0)+a_{23}\frac{\overline{\sigma_z}(\xi,0)H}{E}+a_{24}\frac{\overline{\tau_{rz}}(\xi,0)H}{E}\end{array}\right\} \tag{2.8}$$

where

$$\left.\begin{array}{l}a_{11}=\mathrm{ch}\alpha+\frac{\alpha\mathrm{sh}\alpha}{2(1-\mu)}\\ a_{12}=\frac{1}{2(1-\mu)}[(1-2\mu)\mathrm{sh}\alpha+\alpha\mathrm{ch}\alpha]\\ a_{13}=\frac{1+\mu}{2(1-\mu)}\mathrm{sh}\alpha\\ a_{14}=\frac{1+\mu}{2\alpha(1-\mu)}[(3-4\mu)\mathrm{sh}\alpha+\alpha\mathrm{ch}\alpha]\\ a_{21}=\frac{1}{2(1-\mu)}[(1-2\mu)\mathrm{sh}\alpha-\alpha\mathrm{ch}\alpha]\\ a_{22}=\mathrm{ch}\alpha-\frac{\alpha\mathrm{sh}\alpha}{2(1-\mu)}\\ a_{23}=\frac{1+\mu}{2\alpha(1-\mu)}[(3-4\mu)\mathrm{sh}\alpha-\alpha\mathrm{ch}\alpha]\\ a_{24}=-\frac{1+\mu}{2(1-\mu)}\mathrm{sh}\alpha\end{array}\right\} \tag{2.9}$$

2.2.2 *Calculation Model for Layered Soil Deformation*

If there are aquitards above and below the pumping aquifer, suppose that permeability of the adjacent aquitards is so low that the phreatic stratum groundwater table does not change, then the deformation of upper aquitard and phreatic stratum can be determined if the elastic modulus E and Poisson's ratio μ of the soil layers are known. The elastic modulus E and Poisson's ratio μ of the upper aquitard and phreatic stratum are supposed to be the same. The upper aquitard and the phreatic stratum can be combined to one soil layer, which is named upper soil layer in this chapter. Land subsidence caused by the pumping of fully penetrated confined well can be simplified to a model including single well and two soil layers. One is the aquifer and the another one is named as upper soil layer. Figure 2.2 shows the groundwater table change after pumping. Assumed that the groundwater flow reached a steady state, the groundwater table of the phreatic stratum keeps constant. According to *the principle of effective stress*, pore water pressure of aquifer drops and the effective stress increases. Part of the weight of the upper soil layers is supported by the soil skeleton instead of water. The aquifer undergoes a deformation first for its high permeability. For low permeability, consolidation deformation of the adjacent aquitards within a short time is little. Taking no consideration of the consolidation of upper soil layers, the displacement coordination condition can be adopted. Referring to the idea of sequential coupling model, the confined aquifer deformation is calculated first according to the effective stress change, and then, on the basis of displacement coordination condition, the upper soil layer deformation is studied. The water yield of single-well Q is 3600 m^3/d. The drawdown of pumping well s_w is 15 m, and the unit weight of water is $r_w = 10$ kN/m^3. The parameters of each soil layer are summarized in Table 2.1.

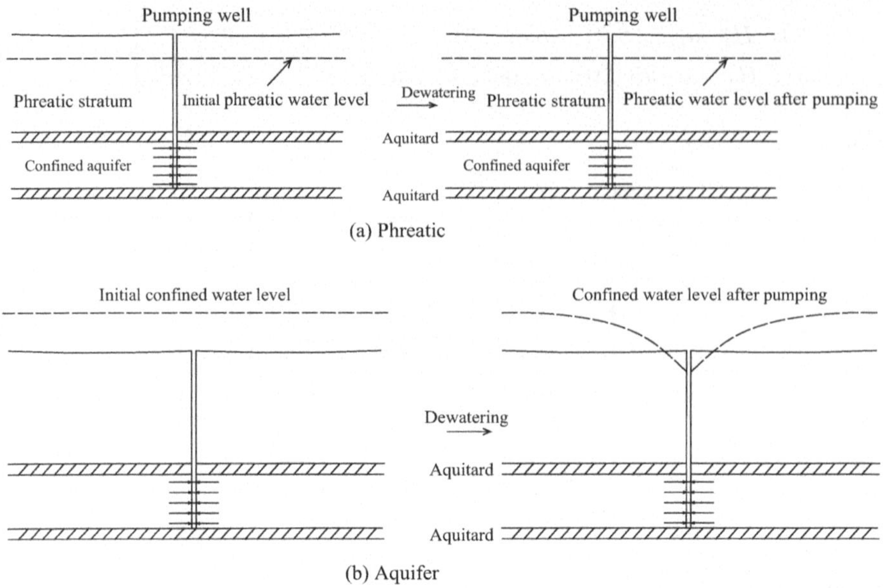

Fig. 2.2 Change of water level

Table 2.1 Calculation of parameters of soil layers

Soil layers	Elastic modulus E (MPa)	Poisson's ratio	Thickness (m)	Permeability coefficient (m/d)
Confined aquifer	20	0.4	20	20
Upper soil layer	4	0.4	30	0

For Dupuit's steady flow of confined well, the drawdown can be described as

$$s_s(r, 0) = \frac{Q}{2\pi T} \ln \frac{R}{r}, \quad (r_0 \leq r \leq R) \tag{2.10}$$

where r is the distance away from the pumping well; r_0 is the radius of pumping well; R is the influence radius, which can be determined by Siechardt's formula:

$$R = 10\, s_w \sqrt{k}, \tag{2.11}$$

where k is the permeability coefficient (m/d).

The pressure acting on confined aquifer induced by weight of upper soil layer does not change. Pore pressure drops and the effective stress increases. It is supposed that the result of effective stress increase equals the consequence that aquifer surface is subjected to the following distribution load (Fig. 2.3):

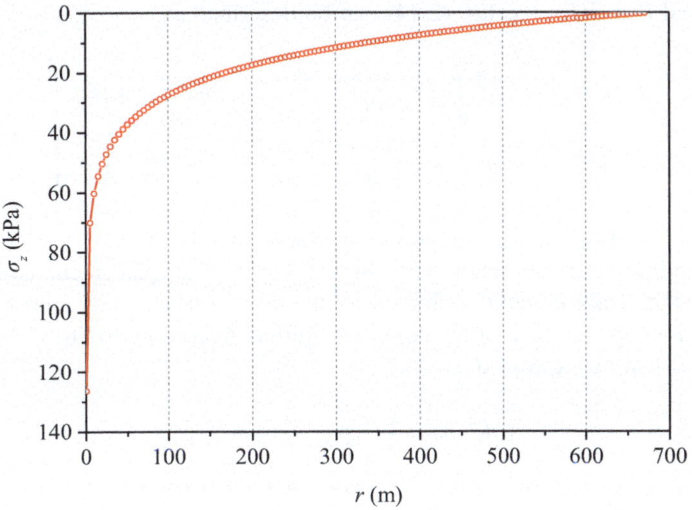

Fig. 2.3 Additional stress acting on aquifer caused by pumping

$$\left.\begin{array}{l}\sigma_z(r,0)=\gamma_w\frac{Q}{2\pi T}\ln\frac{R}{r},\quad (r_0\le r\le R)\\ \tau_{rz}(r,0)=0,\quad (r_0\le r\le R)\end{array}\right\} \tag{2.12}$$

According to Eq. (2.8), if $\overline{\sigma_z}(\xi,H_1)$ and $\overline{\tau_{rz}}(\xi,H_1)$ are known, taking no consideration of the soil layer compression below confined aquifer, that is $\overline{u_r}(\xi,H_1+H_2)=\overline{w}(\xi,H_1+H_2)=0$, $\overline{u_r}(\xi,H_1)$ and $\overline{w}(\xi,H_1)$ can be described as

$$\left.\begin{array}{l}\overline{u_r}(\xi,H_1)=-\frac{a_{22}a_{13}-a_{12}a_{23}}{a_{11}a_{22}-a_{12}a_{21}}\frac{\overline{\sigma_0}H_2}{E_2}\\ \overline{w}(\xi,H_1)=\frac{a_{21}a_{13}-a_{11}a_{23}}{a_{11}a_{22}-a_{12}a_{21}}\frac{\overline{\sigma_0}H_2}{E_2}\end{array}\right\} \tag{2.13}$$

where H_1 and H_2 are the thickness of upper soil layer and confined aquifer, respectively.

The displacement at top surface of confined aquifer is

$$\left.\begin{array}{l}u_r(l,H_1)=-\int_0^\infty\frac{a_{22}a_{13}-a_{12}a_{23}}{a_{11}a_{22}-a_{12}a_{21}}\frac{\overline{\sigma_0}H_2}{E_2}\xi J_1(\xi l)\mathrm{d}\xi\\ w(l,H_1)=\int_0^\infty\frac{a_{21}a_{13}-a_{11}a_{23}}{a_{11}a_{22}-a_{12}a_{21}}\frac{\overline{\sigma_0}H_2}{E_2}\xi J_0(\xi l)\mathrm{d}\xi\end{array}\right\} \tag{2.14}$$

where

$$\overline{\sigma_0}=\int_0^R\sigma_z(r,H_1)rJ_0(r\xi)\mathrm{d}r \tag{2.15}$$

Assumed that $\xi H_2 = \alpha$, Eq. (2.14) can be described as

$$\left.\begin{aligned} u_r(l,H_1) &= \frac{\mu+1}{E_2H_2}\int_0^\infty \frac{(2\mu-1)(4\mu-3)\mathrm{sh}^2\alpha+\alpha^2}{(3-4\mu)\mathrm{sh}^2\alpha+4(1-\mu)^2+\alpha^2}\overline{\sigma_0}J_1\left(\frac{l}{H_2}\alpha\right)\mathrm{d}\alpha \\ w(l,H_1) &= \frac{\mu^2-1}{E_2H_2}\int_0^\infty \frac{(3-4\mu)\mathrm{sh}2\alpha-2\alpha}{(3-4\mu)\mathrm{ch}^2\alpha+(1-2\mu)^2+\alpha^2}\overline{\sigma_0}J_0\left(\frac{l}{H_2}\alpha\right)\mathrm{d}\alpha \end{aligned}\right\} \tag{2.16}$$

There is no load acting on ground surface, that is, $\overline{\sigma_z}(\xi,0) = \overline{\tau_{rz}}(\xi,0) = 0$. $\overline{u_r}(\xi,0)$ and $\overline{w}(\xi,0)$ are unknown. On the basis of displacement coordination condition, the displacement distribution for the lower surface of the upper soil layer equals to the upper surface of the confined aquifer. According to Eq. (2.6), $\overline{u_r}(\xi,0)$ and $\overline{w}(\xi,0)$ can be expressed as

$$\left.\begin{aligned} \overline{u_r}(\xi,0) &= \frac{b_{22}\overline{u_r}(\xi,H_1)-b_{12}\overline{w}(\xi,H_1)}{b_{11}b_{22}-b_{21}b_{12}} \\ \overline{w}(\xi,0) &= \frac{b_{11}\overline{w}(\xi,H_1)-b_{21}\overline{u_r}(\xi,H_1)}{b_{11}b_{22}-b_{21}b_{12}} \end{aligned}\right\} \tag{2.17}$$

Horizontal displacement for upper surface of confined aquifer is much less than the vertical displacement. Taking the calculated result with the parameters in Table 2.1 for example, it is about one order of magnitude smaller (Fig. 2.4). Suppose that the horizontal displacement for upper surface of confined aquifer can be ignored, that is $\overline{u_r}(\xi,H_1) = 0$, Eq. (2.17) can be simplified as

$$\left.\begin{aligned} \overline{u_r}(\xi,0) &= -\frac{b_{12}\overline{w}(\xi,H_1)}{b_{11}b_{22}-b_{21}b_{12}} \\ \overline{w}(\xi,0) &= \frac{b_{11}\overline{w}(\xi,H_1)}{b_{11}b_{22}-b_{21}b_{12}} \end{aligned}\right\} \tag{2.18}$$

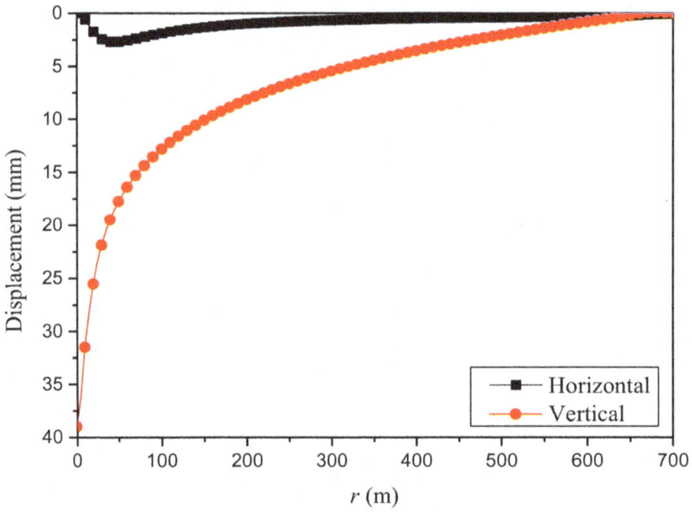

Fig. 2.4 Top surface displacement of confined aquifer

Suppose that $\xi H_1 = \beta$,

$$\left.\begin{aligned} b_{11} &= \operatorname{ch}\beta + \frac{\beta\operatorname{sh}\beta}{2(1-\mu)} \\ b_{12} &= \frac{1}{2(1-\mu)}[(1-2\mu)\operatorname{sh}\beta + \beta\operatorname{ch}\beta] \\ b_{21} &= \frac{1}{2(1-\mu)}[(1-2\mu)\operatorname{sh}\beta - \beta\operatorname{ch}\beta] \\ b_{22} &= \operatorname{ch}\beta - \frac{\beta\operatorname{sh}\beta}{2(1-\mu)} \end{aligned}\right\} \tag{2.19}$$

Substitute Eqs. (2.13)–(2.18),

$$\left.\begin{aligned} \overline{u_r}(\xi,0) &= -\frac{b_{12}}{b_{11}b_{22}-b_{21}b_{12}} \times \frac{a_{21}a_{13}-a_{11}a_{23}}{a_{11}a_{22}-a_{12}a_{21}}\frac{\overline{\sigma_0}H_2}{E_2} \\ \overline{w}(\xi,0) &= \frac{b_{11}}{b_{11}b_{22}-b_{21}b_{12}} \times \frac{a_{21}a_{13}-a_{11}a_{23}}{a_{11}a_{22}-a_{12}a_{21}}\frac{\overline{\sigma_0}H_2}{E_2} \end{aligned}\right\} \tag{2.20}$$

Inverse transform $\overline{u_r}(\xi,0)$ and $\overline{w}(\xi,0)$, and the ground surface deformation can be obtained.

$$\left.\begin{aligned} u_r(l,0) &= -\int_0^\infty \frac{b_{12}}{b_{11}b_{22}-b_{21}b_{12}} \times \frac{a_{21}a_{13}-a_{11}a_{23}}{a_{11}a_{22}-a_{12}a_{21}}\frac{\overline{\sigma_0}H_2}{E_2}\xi J_1(\xi l)\mathrm{d}\xi \\ w(l,0) &= \int_0^\infty \frac{b_{11}}{b_{11}b_{22}-b_{21}b_{12}} \times \frac{a_{21}a_{13}-a_{11}a_{23}}{a_{11}a_{22}-a_{12}a_{21}}\frac{\overline{\sigma_0}H_2}{E_2}\xi J_0(\xi l)\mathrm{d}\xi \end{aligned}\right\} \tag{2.21}$$

where

$$\overline{\sigma_0} = \int_0^R \sigma_z(r, H_1) r J_0(r\xi)\mathrm{d}r \tag{2.22}$$

2.2.3 *Case Study*

2.2.3.1 Soil Deformation Caused by Steady Flow of Confined Well

Equation (2.21) tells that the ground surface deformation has no relationship with elastic modulus of upper soil layer. However, the parameter was adopted to calculate land subsidence in the papers using stress coordination condition (Gong and Zhang 2011; Wang et al. 2013). To verify Eq. (2.21), numerical simulation was conducted with the parameters shown in Table 2.1. Suppose that water inflow Q of well is 3600 m^3/d. The drawdown of pumping well s_w is 15 m, and the unit weight of water r_w is 10 kN/m^3. Figure 2.5 shows the simulation model. Pore pressure of confined aquifer at 700 m apart from the pumping well was fixed. Suppose that the horizontal displacement of confined aquifer at the top surface is zero.

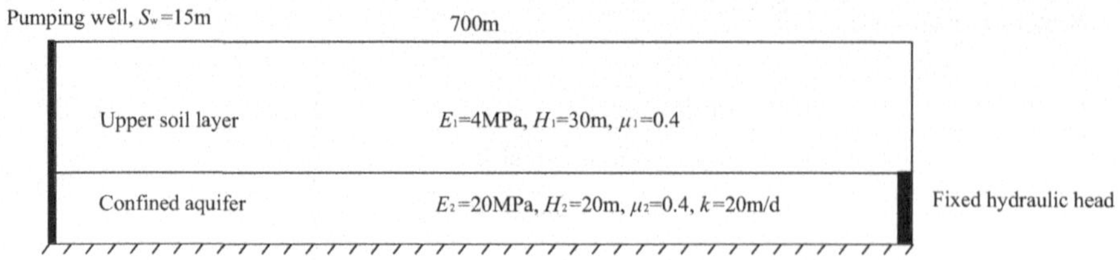

Fig. 2.5 Numerical simulation model

Figure 2.6 shows that with the modulus of upper soil layer changing from 4 to 40 MPa, the maximum change of horizontal and vertical ground surface displacement is less than 5%, meaning that the ground surface displacement has little relationship with the modulus of upper soil layer change.

If the thickness change of upper soil layer affects the calculated ground surface deformation little either, then the upper soil layer deformation may be ignored. Figure 2.7 shows that when the thickness is 30 m, the maximum settlement is 31.3 mm. Assumed that the thickness increased to 60, 90, and 120 m, the maximum settlement decreased to 22.7, 32.2, and 39.3%. The ground surface subsidence of the position within 50 m away from the pumping well changes distinctly. Although the maximum horizontal displacement keeps at 3.3 mm, the maximum displacement position moved from 40 to 75, 105, and 135 m, respectively. The horizontal displacement distribution changes significantly. Compared with changing elastic modulus, change of upper soil layer thickness has a greater impact on ground surface deformation.

Suppose that drawdown of pumping well s_w is 15 m and the unit weight of water r_w is 10 kN/m^3. The parameters in Table 2.1 and Fig. 2.8 show the vertical displacement at confined aquifer and ground surface calculated with Eqs. (2.16) and (2.21). Within 40 m away from the pumping well, the difference between displacements at confined aquifer and ground surface is distinct. At the pumping point, the ground surface settlement is 26.7 mm, which is only 70% to that of confined aquifer. This means the existence of upper soil layer can mitigate land subsidence. Similar conclusion can be found in numerical simulations and in situ tests (Wang et al. 2009; Zheng et al. 2014). The settlement difference between confined aquifer surface and ground surface reduced with the distance r increasing. Zhou et al. (2010, 2011) considered the compression of aquifer to be equal to land subsidence. From the analysis above, it can be told that this method tends to be larger.

To verify the result with proposed method, corresponding numerical simulation was conducted. Comparing the theoretical result with numerical simulation, the maximum deviation for aquifer surface settlement is about 15% and the position is located at about 50 m away from the pumping point. The maximum deviation for ground surface settlement is about 15% and the position is about 50–100 m away from the pumping point. Deviation from other positions is less than 10% (Fig. 2.9). The maximum horizontal displacement is 3.3 mm, located at 40 m away from the

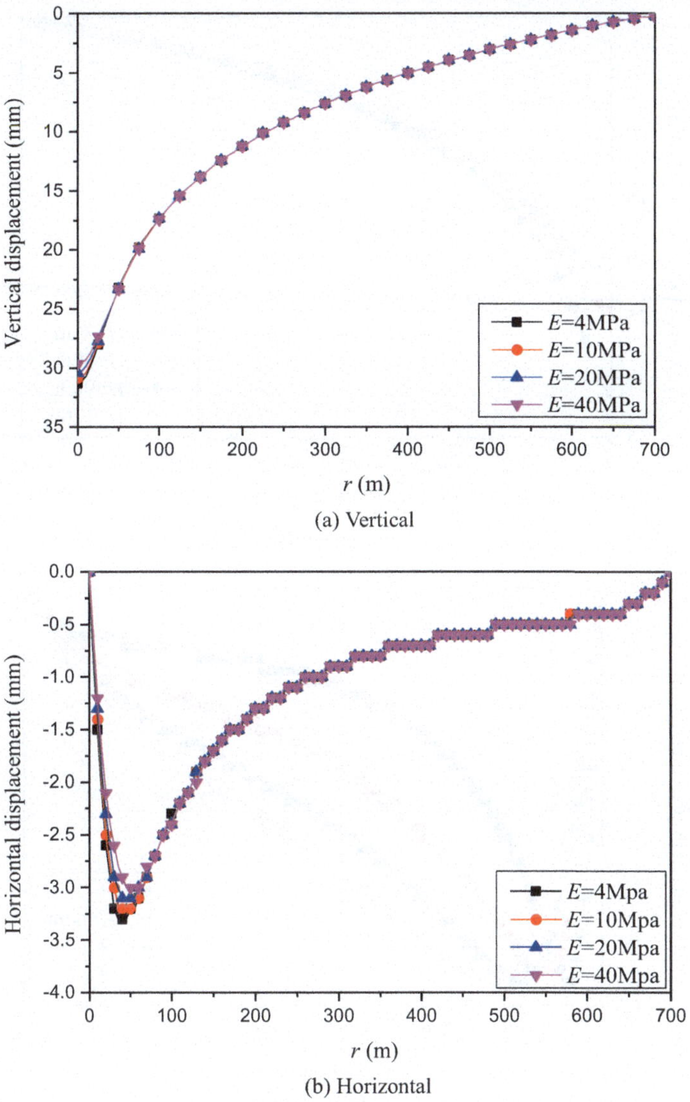

Fig. 2.6 Ground surface displacement of a different upper soil layer elastic modulus

pumping point. The theoretical result of ground surface horizontal displacement is close to numerical simulation result (Fig. 2.10).

2.2.3.2 Soil Deformation Caused by Transient Flow of Confined Well

For unsteady flow confined well, aquifer drawdown can be described as

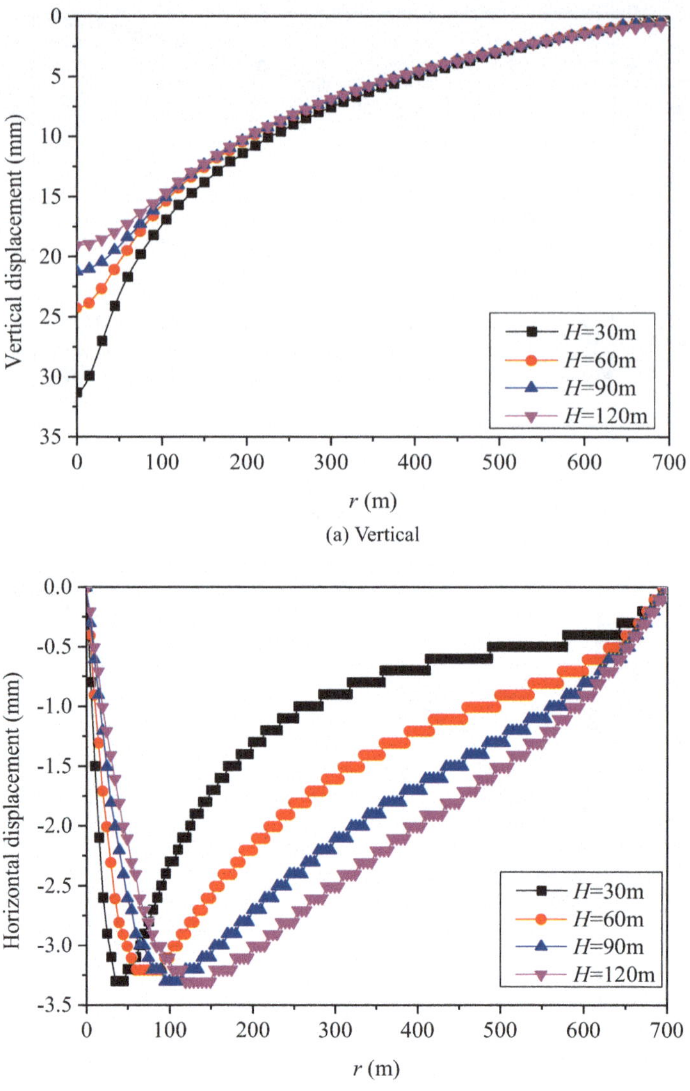

(a) Vertical

(b) Horizontal

Fig. 2.7 Ground surface displacement for different upper soil layer thicknesses

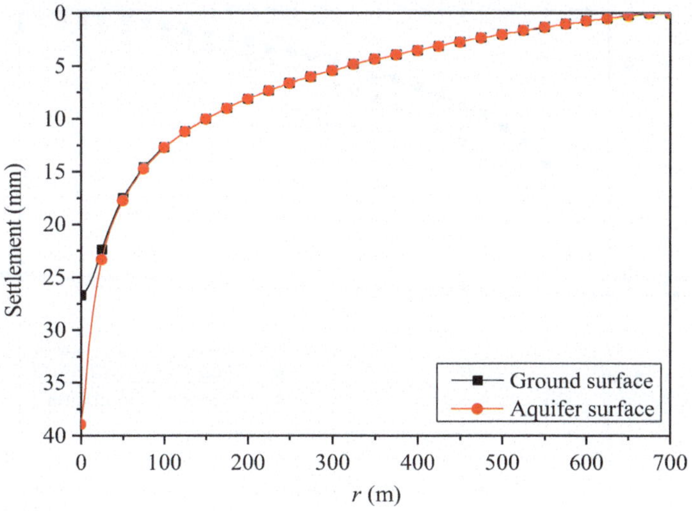

Fig. 2.8 Settlement of ground and aquifer surface

$$\left.\begin{aligned} s(r,t) &= \frac{Q}{4\pi KH} W(u) \\ W(u) &= \int_u^{\infty} \frac{e^{-y}}{y} dy \\ u &= \frac{\gamma_w(\alpha + n\beta)r^2}{4Kt} \end{aligned}\right\} \tag{2.23}$$

where α is the soil skeleton volume compressibility; β being the water volume compressibility; and n being the void ratio.

Suppose that u is small enough, $W(u)$ and $s(r,t)$ can be approximately described as

$$\left.\begin{aligned} W(u) &\approx \ln \frac{2.25Kt}{\gamma_w(\alpha + n\beta)r^2} \\ s(r,t) &= \frac{Q}{4\pi KH} \times \ln \frac{2.25Kt}{\gamma_w(\alpha + n\beta)r^2} \end{aligned}\right\} \tag{2.24}$$

Assume that water is incompressible, that is, $\beta = 0$. $\sigma_z(r,0)$ can be described as

$$\sigma_z(r,0) = \frac{\gamma_w Q}{4\pi KH} \times \ln \frac{2.25Kt}{\gamma_w \alpha r^2} \tag{2.25}$$

Figure 2.11 shows the displacement of confined aquifer and upper soil layer surface. Settlements keep growing with pumping time increasing. Pumping for 1, 5, 10, and 30 days, the maximum settlement at ground surface, located at pumping point, increases from 21.3 to 26.7, 29.0, and 32.7 mm, which is 63.5, 68.6, 70.4, and 72.8% of the maximum confined aquifer surface settlement, respectively. The settlement region also expands. Pumping for 1 day, settlement of position more than

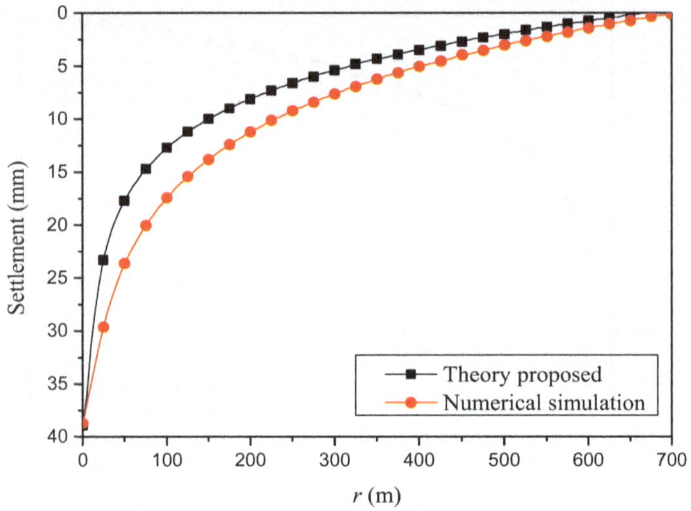

(a) Aquifer surface

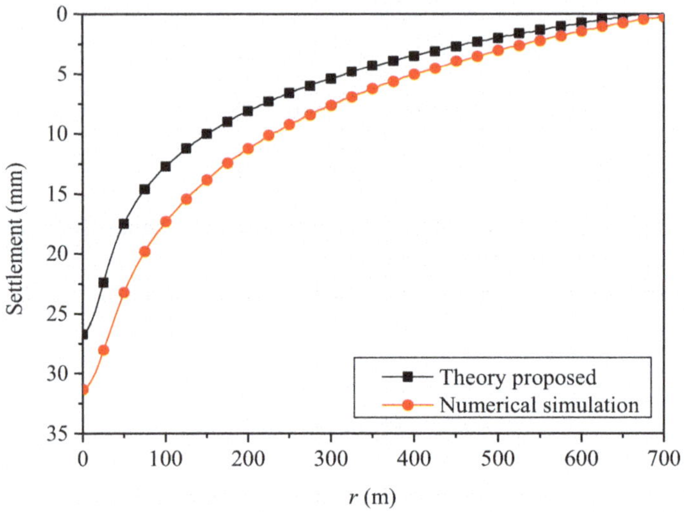

(b) Ground surface

Fig. 2.9 Comparison of numerical modeling with theoretical analysis for vertical displacement

300 m away from pumping point is zero, while Pumping for 5 days, the distance increased to 670 m. Pumping for 10 days and 30 days, settlements of position at 700 m away from pumping point are 2.0 and 5.7 mm, respectively. The settlement difference between confined aquifer and upper soil layer surface is distinct within 40 m away from the pumping well. Located at 40 m away from pumping point, the maximum horizontal displacement at ground surface keeps at 3.3 mm with

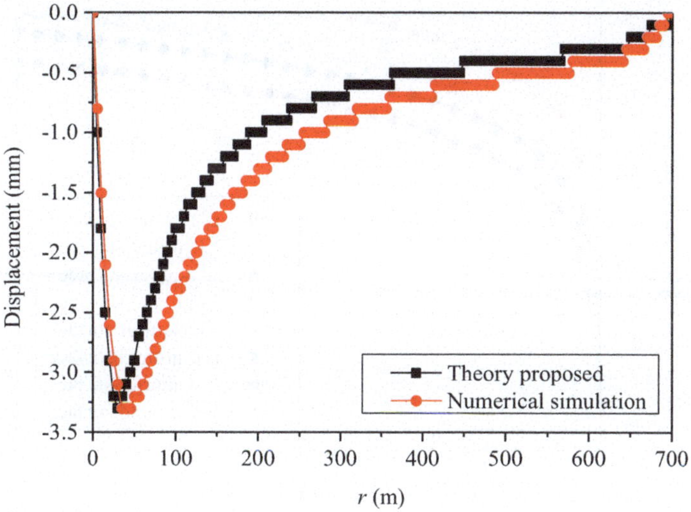

Fig. 2.10 Comparison of numerical modeling with theoretical analysis for horizontal displacement

pumping time increasing. The horizontal displacement range expands. Pumping for 1 day, horizontal displacement at position more than 350 m away from pumping point is zero, while pumping for 5 days, the distance increased to 700 m. Pumping for 10 and 30 days, horizontal displacements of position at 700 m away from pumping point are both 0.3 mm. After pumping for a certain time, the horizontal displacement does not change with time.

2.2.3.3 Soil Deformation Caused by In Situ Pumping Test

The in situ pumping test at Cultural Square Station of Tianjin Metro Lines 5 and 6 is taken as an example to verify the proposed theory. The pumping test drawdown $s_w = 12$ m. The test lasted 2 days. Table 2.2 shows the soil physical parameters (Zheng et al. 2014). The confined aquifer is comprised of three soil layers. The adopted elastic modulus $E = 82.3$ MPa, which is equal to weighted average of three soil layers. Poisson's ratio $\mu = 0.3$. For water in aquifer flows along horizontal direction, the aquifer permeability coefficient is determined by weighted average of horizontal permeability coefficient of three soil layers, which is equal to 2 m/day. The influence radius R equals to 180 m. The pumping well radius $r_0 = 0.14$ m and unit weight of water $r_w = 10$ kN/m^3.

The critical step of the calculation model for layered soil deformation is to determine the load distribution acting on aquifer surface. For a pumping well, the relationship between discharge and drawdown is

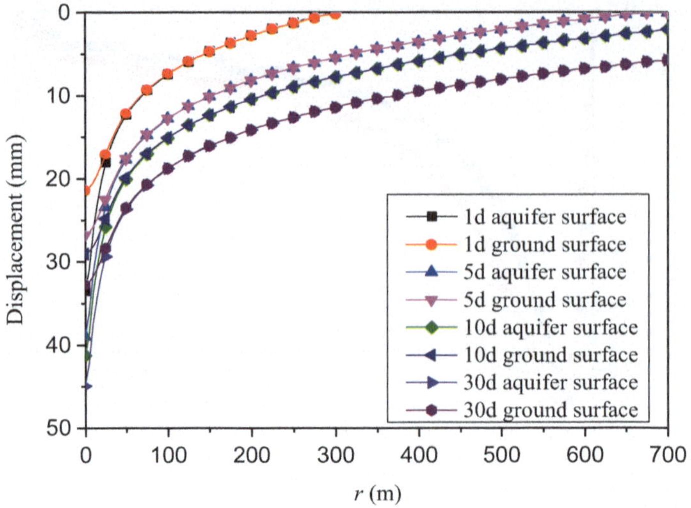

(a) Vertical

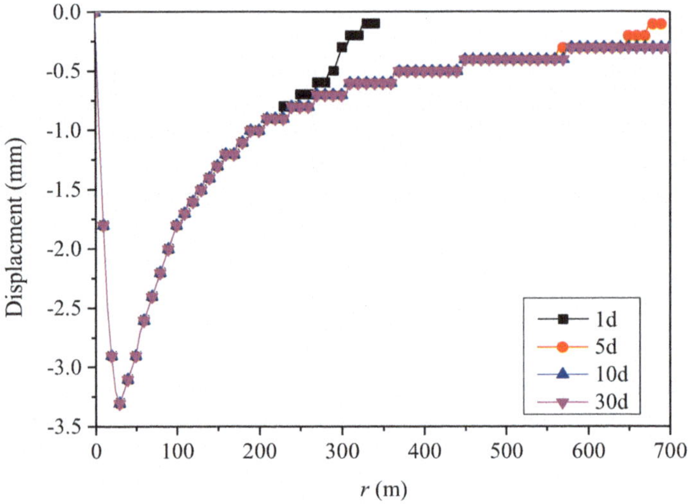

(b) Horizontal

Fig. 2.11 Displacement of aquifer and ground surface at different pumping times

Table 2.2 Soil parameters for cultural square station

Soil strata	Soil	Buried depth of soil layer (m)	Elastic modulus E (kPa)	Poisson's ratio μ	Horizontal permeability coefficient k_x (m/day)	Vertical permeability coefficient k_z (m/day)
Overlying soil layers	Silty clay	16.5	29,778	0.30	0.002	0.0003
The first aquifer	Silty sand	20.5	94,358	0.25	3.000	1.0000
	Silty clay	23.5	43,835	0.30	0.005	0.0023
	Silty sand	29.0	94,358	0.25	2.210	1.5200

$$Q' = \frac{2\pi T s_w}{\ln\left(\frac{R}{r_0}\right)} \tag{2.26}$$

Load acting on aquifer surface is

$$\sigma_z(r,0) = \gamma_w \frac{Q'}{2\pi T} \ln\left(\frac{R}{r}\right) = \gamma_w s_w \frac{\ln\left(\frac{R}{r}\right)}{\ln\left(\frac{R}{r_0}\right)} \tag{2.27}$$

To protect measuring device, the measured points were buried 2 m beneath ground surface. Figure 2.12 shows that within 20 m from pumping well, difference between subsidence at aquifer surface and 2 m beneath ground surface was distinct. Near the pumping well, subsidence at aquifer surface was 7.9 mm, while 5.9 mm at the measured point. The latter was about 74% of the former. The maximum horizontal displacement of ground surface was 0.8 mm, located at about 20 m from the pumping well.

Figure 2.13 shows the comparison of ground subsidence obtained by the proposed theory, numerical simulation, and in situ measurement. Measured subsidence at DCCJ07, DCCJ06, DCCJ04, and DCCJ01 are 4.8, 4.3, 3.4, and 2.3 mm, respectively. The numerical simulation subsidence is 5.1, 4.7, 4.3, and 3.6 mm, respectively. Subsidences obtained by the proposed theory are 5.6, 5.0, 4.2, and 3.3 mm, respectively. Subsidence obtained by the proposed theory approximates with subsidence obtained by numerical simulation and in situ test.

Figure 2.14 shows the comparison of calculated subsidence with monitoring data of soil settlement around pumping well. Along measured points FCCJ1 and FCCJ2, which are near the pumping well, soil subsidence decreases markedly with the buried depth reducing. Along measured points FCCJ3 and FCCJ4, which are a certain distance away from pumping well, soil subsidence does not change with buried depth.

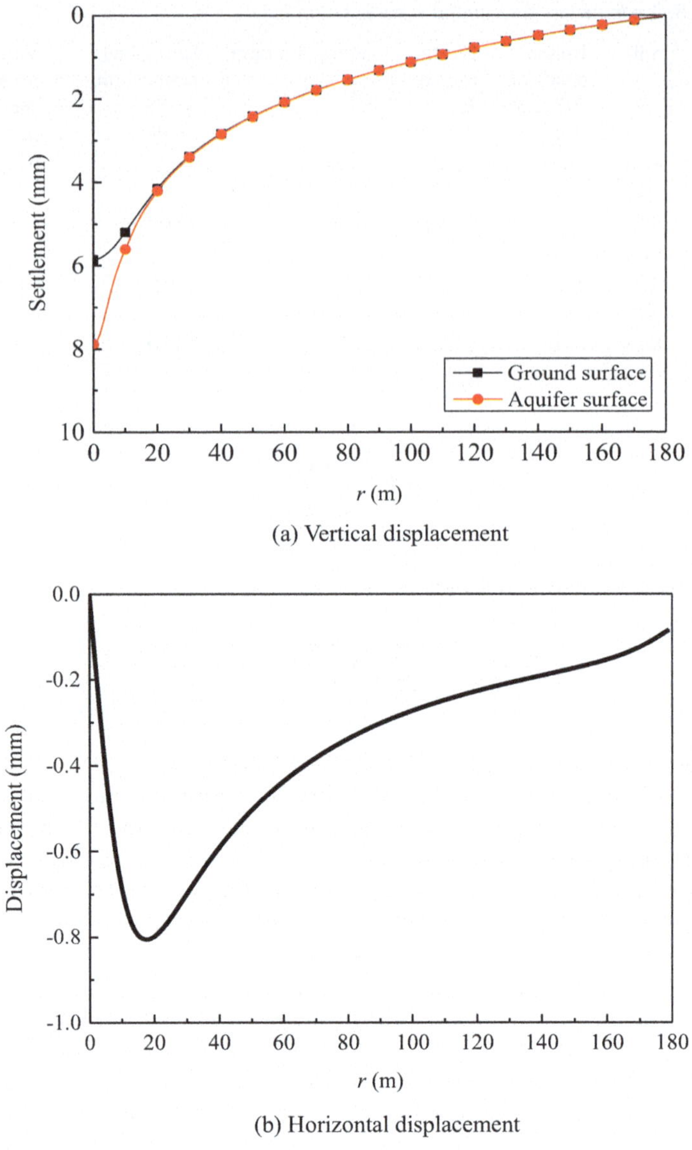

Fig. 2.12 Deformation of confined aquifer and ground surface

2.3 Land Subsidence Caused by High-Rise Building Load

The subsidence caused by group piles is affected by many factors, including the geometry size of the group piles, the pile foundation construction, the physical and mechanics characteristics of the soil, the change of the soil layer, the load magnitude

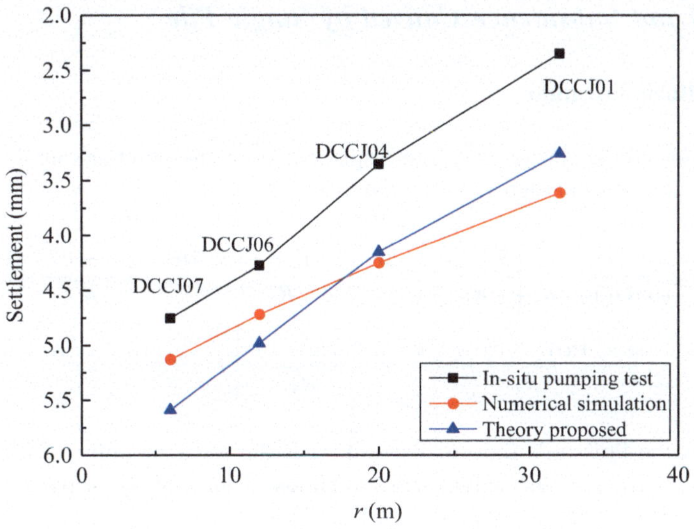

Fig. 2.13 Ground settlement around pumping well

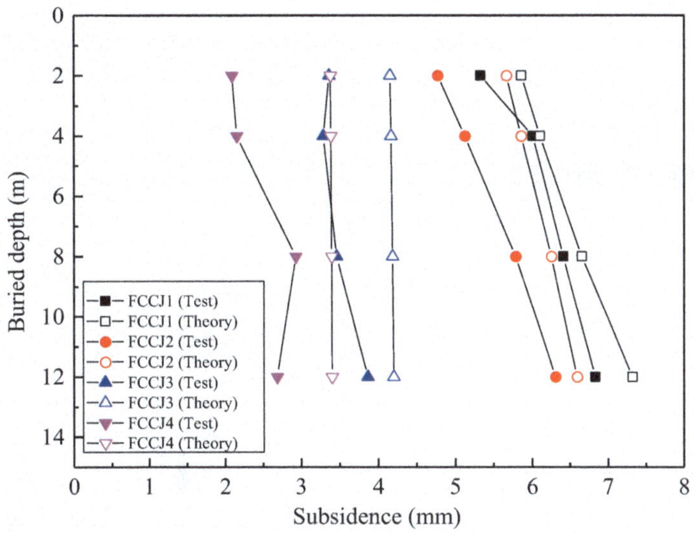

Fig. 2.14 Comparison of calculated results with monitoring data of soil settlement

and its operation time, and so on. It is difficult to put forward the method of computing subsidence caused by group piles considering the above factors (Gong 2001).

Based on the stress of Geddes solution (1966), the formula of computing subsidence caused by the single pile was derived by Mindlin displacement solution (Cui 2012a).

2.3.1 *Land Subsidence Caused by Single Pile*

2.3.1.1 Basic Formula

To the semi-infinite elastic body, the displacement of any point among the uniform soil under the concentrated force is given by

$$w = \frac{Q}{16\pi G(1-\mu)}\left[\frac{3-4\mu}{R_1} + \frac{8(1-\mu)^2-(3-4\mu)}{R_2} + \frac{(z-c)^2}{R_1^3} + \frac{(3-4\mu)(z+c)^2-2cz}{R_2^3} + \frac{6cz(z+c)^2}{R_2^5}\right] \tag{2.28}$$

where μ is the Poisson's ratio of the soil; G is the shear modulus of the soil and $G = E/2(1+\mu)$; E is the deformation modulus of the soil; Q is the concentrated load; and other variables are shown in Fig. 2.15.

2.3.1.2 Subsidence Induced by the Pile-Tip Concentrated Load

Using Formula (2.28), the subsidence induced by the pile-tip concentrated load was obtained:

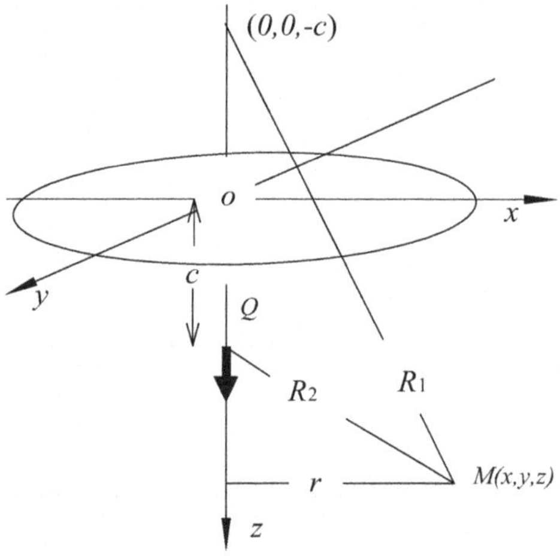

Fig. 2.15 Force diagram

$$
\begin{aligned}
w &= \frac{P_b}{16\pi G(1-\mu)}\left[\frac{3-4\mu}{R_1} + \frac{8(1-\mu)^2-(3-4\mu)}{R_2} + \frac{(z-l)^2}{R_1^3}\right. \\
&\quad \left. + \frac{(3-4\mu)(z+l)^2-2lz}{R_2^3} + \frac{6lz(z+l)^2}{R_2^5}\right] \\
&= \frac{P_b}{16\pi Gl(1-\mu)}\left[\frac{3-4\mu}{A} + \frac{8(1-\mu)^2-(3-4\mu)}{B} + \frac{(m-1)^2}{A^3}\right. \\
&\quad \left. + \frac{(3-4\mu)(m+1)^2-2m}{B^3} + \frac{6m(m+1)^2}{B^5}\right] = \frac{P_b}{l}I_b
\end{aligned} \tag{2.29}
$$

$$
\begin{aligned}
I_b &= \frac{P_b}{16\pi G(1-\mu)}\left[\frac{3-4\mu}{A} + \frac{8(1-\mu)^2-(3-4\mu)}{B} + \frac{(m-1)^2}{A^3}\right. \\
&\quad \left. + \frac{(3-4\mu)(m+1)^2-2m}{B^3} + \frac{6m(m+1)^2}{B^5}\right]
\end{aligned} \tag{2.30}
$$

where $A^2 = n^2 + (m-1)^2$; $B^2 = n^2 + (m+1)^2$; $F^2 = n^2 + m^2$; $m = z/l$; $n = r/l$.

2.3.1.3 Subsidence Induced by Different Lateral Frictions of Pile

(1) Rectangular distribution of the lateral friction along the pile

Figure 2.16a illustrates rectangular distribution of the lateral friction along the pile. The resultant force of the lateral friction of pile is P, and the load for $\mathrm{d}h$ is $\mathrm{d}q = (P/l)\mathrm{d}h$. The vertical subsidence of any point $A(r, z)$ among the soil is

$$
\begin{aligned}
w &= \int_0^l \frac{\mathrm{d}q}{16\pi G(1-\mu)}\left[\frac{3-4\mu}{R_1} + \frac{8(1-\mu)^2-(3-4\mu)}{R_2} + \frac{(z-h)^2}{R_1^3}\right. \\
&\quad \left. + \frac{(3-4\mu)(z+h)^2-2hz}{R_2^3} + \frac{6hz(z+h)^2}{R_2^5}\right] \\
&= \int_0^l \frac{P\mathrm{d}h}{16\pi Gl(1-\mu)}\left[\frac{3-4\mu}{R_1} + \frac{8(1-\mu)^2-(3-4\mu)}{R_2} + \frac{(z-h)^2}{R_1^3}\right. \\
&\quad \left. + \frac{(3-4\mu)(z+h)^2-2hz}{R_2^3} + \frac{6hz(z+h)^2}{R_2^5}\right] = \frac{P}{l}I_r,
\end{aligned} \tag{2.31}
$$

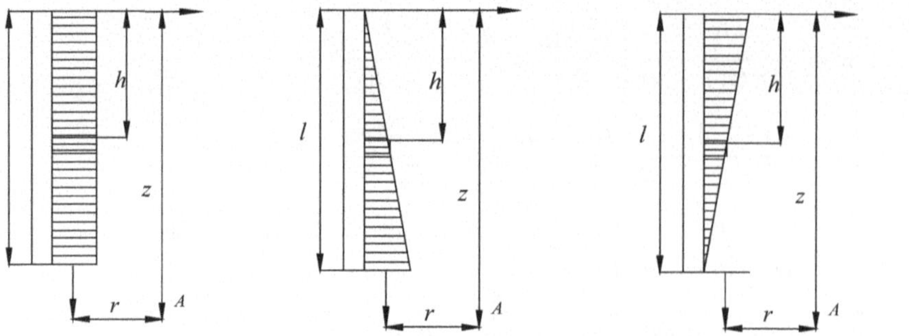

(a) Rectangular distribution (b) Erect triangular distribution (c) Inverted triangular distribution

Fig. 2.16 Distribution of lateral friction of pile

where

$$I_r = \frac{1}{16\pi G(1-\mu)}\left[4(1-\mu)\ln\frac{1-m+A}{F-m} + 8(1-\mu)^2\ln\frac{1+m+B}{F+m} + \frac{m-1}{A} - \frac{(3-\mu)(m+1)+6m}{B} + \frac{2mn^2}{B^3} + \frac{4(1-\mu)m}{F} - \frac{2m^2(m+1)^3}{n^2B^3} + \frac{2m(n^2+m+m^2)}{n^2B}\right]$$

(2) Erect triangular distribution of the lateral friction along the pile

Figure 2.16b illustrates the erect triangular distribution of the lateral friction along the pile. The resultant force of the lateral friction of pile is P. The load at the bottom margin of the triangle is $\frac{2P}{l}$ and the load for dh is $dq = \frac{2P}{l^2}hdh$.

The vertical subsidence of any point $A(r, z)$ among the soil is

$$\begin{aligned} w &= \int_0^l \frac{dq}{16\pi G(1-\mu)}\left[\frac{3-4\mu}{R_1} + \frac{8(1-\mu)^2-(3-4\mu)}{R_2} + \frac{(z-h)^2}{R_1^3} + \frac{(3-4\mu)(z+h)^2-2hz}{R_2^3} + \frac{6hz(z+h)^2}{R_2^5}\right] \\ &= \int_0^l \frac{Pdh}{8\pi Gl^2(1-\mu)}\left[\frac{3-4\mu}{R_1}h + \frac{8(1-\mu)^2-(3-4\mu)}{R_2}h + \frac{(z-h)^2}{R_1^3}h + \frac{(3-4\mu)(z+h)^2-2hz}{R_2^3}h + \frac{6hz(z+h)^2}{R_2^5}h\right] = \frac{P}{l}I_t, \end{aligned} \tag{2.32}$$

where

$$I_t = \frac{1}{8\pi G(1-\mu)}\left\{(5-4\mu)A + \frac{m-1}{A} + \left[8(1-\mu)^2 + (3-4\mu)\right]B \right.$$

$$- 4m(1-\mu)\ln\frac{F-m}{A-m+1} + \left[8(1-\mu)^2 - 4\right]m\ln\frac{F+m}{B+m+1}$$

$$\left. + \frac{2m(m+1)^2\left[m^2(m+1) - n^2(m-1)\right]}{n^2B^3}\right\}$$

(3) Inverted triangular distribution of the lateral friction along the pile

Figure 2.16c illustrates the inverted triangular distribution of the lateral friction along the pile, which can be regarded as the difference of the rectangular distribution and the erect triangular distribution of the lateral friction. Assume that the total resultant force of the lateral friction of the pile is P. The resultant force of the rectangular distribution is $2P$ and that of the erect triangular distribution is P. The subsidence induced by the inverted triangular distribution of the lateral friction along the pile is

$$w = \frac{2P}{l}I_r - \frac{P}{l}I_t \tag{2.33}$$

(4) Erect trapezoidal distribution of the lateral friction along the pile

This can be regarded as the sum of the rectangular distribution and the erect triangular distribution of the lateral friction. Assume the ratio of the rectangle to the erect trapezoid is β, so the resultant force of the rectangular distribution is βP and that of the erect triangular distribution is $(1-\beta)P$. The subsidence induced by the erect trapezoidal distribution of the lateral friction along the pile is

$$w = \frac{\beta P}{l}I_r + \frac{(1-\beta)P}{l}I_t \tag{2.34}$$

(5) Inverted trapezoidal distribution of the lateral friction along the pile

This can be regarded as the difference of the rectangular distribution and the erect triangular distribution of the lateral friction. Assume the ratio of the rectangle to the erect trapezoid is β, so the resultant force of the rectangular distribution is βP and that of the erect triangular distribution is $(1-\beta)P$. The subsidence induced by the inverted trapezoidal distribution of the lateral friction along the pile is

$$w = \frac{\beta P}{l} I_r - \frac{(1-\beta)P}{l} I_t \tag{2.35}$$

(6) Other distributions of the lateral friction along the pile

This can be simplified to be the combination of several rectangles and triangles, which can be calculated by the above formulas, respectively.

2.3.2 *Land Subsidence Caused by Pile Group*

According to the above theory, the influence of any concentrated load in the elastic body is infinite and the influence of the pile foundation on the surrounding soil is also infinite in theory. If the number of piles is small and the distance among piles is large, the additional pressure from the piles at the end of the pile plane does not overlap or overlap little when the pile group is loaded. The working state of each pile in the pile group is similar to that of the single pile. However, if the number of piles is larger and the distance among piles is small, the additional pressure at the end of the pile plane will be the superposition of each pile. Because of the additional stress increasing and the influence range widening and deepening, the settlement of pile groups is much higher than that of a single pile. Especially, when there is a high compressible soil layer under the pile bearing stratum, the settlement will further increase under the stress superposition of the pile foundation.

2.4 Subsidence Caused by Pumping and High-Rise Buildings

The variations of groundwater level and the building load are two independent factors that cause the land subsidence by changing the effective stress of soil. Previous studies have shown that there is a coupling effect between the two factors and the superposition value is less than the sum of the values produced by the two individual factors (Ding et al. 2011).

Considering the interaction of piles and soils, the soil deformation caused by the decline of groundwater level may change the stress state of the pile and the pile stress state change in turn prevents the subsidence of soil, so the pile foundation is conducive to reduce ground settlement. In fact, the ground subsidence is only the appearance; it is caused by the consolidation of the soil for either the building load or the groundwater level change.

2.5 Conclusions

When decompression of confined aquifer features short duration, consolidation of confined aquifer can complete within a short time for its high permeability. And consolidation deformation of the adjacent aquitards within a short time is little for its low permeability. The aquitards are supposed to be watertight. Based on displacement coordination condition, theory for elastic-layered system is adopted to calculate layered soil deformation caused by decompression of confined aquifer, including vertical and horizontal components. The calculated results are compared with numerical simulation and in situ pumping test. It is necessary to point out that decompression of only one aquifer in subsidence model is built in this chapter. If decompression of more than one aquifer was conducted simultaneously, the final displacement can be obtained by adding the calculated deformation caused by decompression of each aquifer. The following conclusion can be drawn:

(1) The elastic modulus of overlying soil has little influence on ground surface displacement caused by decompression of confined water, yet thickness of overlying soil should be taken into consideration.
(2) Settlement of soils overlying confined aquifer increases up to down nearby the pumping well. Over a certain distance, the settlement does not change with buried depth.
(3) For unsteady flow confined wells, the settlement value and region at ground and aquifer surface grows with pumping time increasing. The maximum horizontal displacement value and position at ground surface does not change with time after pumping for a certain time.

References

Burmister DM (1944) The general theory of stresses and displacements in layer systems. I J Appl Phys 16:89–94

Burmister DM (1945a) The general theory of stresses and displacements in layered soil systems. II J Appl Phys 16:126–127

Burmister DM (1945b) The general theory of stresses and displacements in layered soil systems. III J Appl Phys 16:296–302

Chen XX, Luo ZJ, An XY, Tan JZ, Tian KY (2013) Coupling model of groundwater three dimensional variable-parametric non-steady seepage and land subsidence. J Jilin Univ (Earth Sci Ed) 43(5):1572–1578 (in Chinese)

Cui ZD (2012a) Bearing capacity of single pile and in-flight T-bar penetration for centrifuge modeling of land subsidence caused by the interaction of high-rise buildings. Bull Eng Geol Environ 71:579–586

Cui ZD (2012b) Land subsidence disaster caused by natural factors and human activities. Disaster Adv 5(2):3 4

Cui ZD, Jia YJ, Yuan L (2016) Distribution law of soil deformation caused by decompression of confined water. Environ Earth Sci 75:1281

Ding DM, Ma FS, Zhang YM, Wang J, Guo D (2011) Characteristics of land subsidence due to both high-rise building and exploitation of groundwater in urban area. J Eng Geol 19(3):433–439
Galloway DL, Burbey TJ (2011) Review: regional land subsidence accompanying groundwater extraction. Hydrogeol J 19(8):1459–1486
Geddes JD (1966) Stresses in foundation soils due to vertical subsidence load. Geotechnique 16 (3):231–255
Gong QM (2001) Settlement calculation based on Mindlin displacement solution for pile group. Undergr Space 21(3):167–172
Gong XN, Zhang J (2011) Settlement of overlying soil caused by decompression of confined water. Chin J Geotech Eng 33(1):145–149 (in Chinese)
Lou RX, Zhou NQ, Zhao S (2011) Numerical simulation of deep foundation pit dewatering of Xujiahui station of Shanghai Metro Line No. 11. Chin J Undergr Space Eng 7(5):908–913 (in Chinese)
Luo GY, Pan H, Chao H, Yin XL (2004) Analysis of settlements caused by decompression of confined water. Rock Soil Mech 25(S2):196–200 (in Chinese)
Luo ZJ, Liu JB, Li L (2008) Three-dimensional full coupling numerical simulation of groundwater dewatering and land-subsidence in quaternary loose sediments. Chin J Geotech Eng 30(2):193–198 (in Chinese)
Luo ZJ, Li L, Yao TQ, Luo JJ (2006) Coupling model of three dimensional seepage and land-subsidence for dewatering of deep foundation pit in loose confined aquifers. Chin J Geotech Eng 28(11):1947–1951 (in Chinese)
Luo ZJ, Wang Y, Tian XW, Tian JH (2013) Simulating and forecasting of groundwater exploitation, land subsidence and ground fissure in Cangzhou City. Chin J Hydraul Eng 44 (2):198–204 (in Chinese)
Miu JF, Wu LG, Wang ZQ (1991) Land subsidence caused by large-scale dewatering from deep pumping wells. Chin J Geotech Eng 13(3):60–64 (in Chinese)
Shen SL, Xu YS (2011) Numerical evaluation of land subsidence induced by groundwater pumping in Shanghai. Can Geotech J 48(9):1378–1392
Wang CB, Ding WQ, Liu WJ, Qiao YF (2013) Distribution law of soil settlement caused by unsteady dewatering of confined water. J Tongji Univ (Nat Sci) 41(3):361–367 (in Chinese)
Wang JX, Wu LG, Zhu YF, Tang YQ, Yang P, Lou RX (2009) Mechanism of dewatering-induced ground subsidence in deep subway station pit and calculation method. Chin J Rock Mechan Eng 28(5):1010–1018 (in Chinese)
Wu H, Xie KH, Huang DZ (2014) Analytical solution for one-dimensional consolidation of structured aquitard soils in second kind of leakage system. Chin J Geotech Eng 36(9):1688–1695 (in Chinese)
Wu JC, Shi XQ, Ye SJ, Xue YQ, Zhang Y, Wei ZX, Fang Z (2010) Numerical simulation of viscoelastoplastic land subsidence due to groundwater overdrafting in Shanghai, China. J Hydrol Eng 15(3):223–236
Zeng J, Wang XD (2012) Analytical solution to one-dimensional consolidation of aquitards for unsteady flow in confined aquifers. J Nanjing Univ Technol (Nat Sci Ed) 34(2):85–90 (in Chinese)
Zhang ZM, Zhao GH (1985) Computing displacements and stresses of multilayer soils subjected to axis-symmetric loading by initial function method. J East China Tech Univ Water Resour 13 (4):40–48 (in Chinese)
Zhang ZM, Zhao GH (1986) Computing displacements and stresses of transversely isotroic layers by initial function method. J East China Tech Univ Water Resour 14(3):82–89 (in Chinese)
Zheng G, Zeng CF, Xue XL (2014) Settlement mechanism of soils induced by local pressure-relief of confined aquifer and parameter analysis. Chin J Geotech Eng 36(5):802–817 (in Chinese)
Zhou NQ, Pieter A, Vermeer Lou RX, Tang YQ, Jiang SM (2010) Numerical simulation of deep foundation pit dewatering and optimization of controlling land subsidence. Eng Geol 114:251–260
Zhou NQ, Tang YQ, Lou RX, Jiang SM (2011) Numerical simulation of deep foundation pit dewatering and land subsidence control of Xujiahui Metro Station. Chin J Geotech Eng 33 (12):1950–1956 (in Chinese)

Chapter 3
Consolidation of Saturated Multilayered Soils Caused by Pumping from the Dewatering Well

3.1 Introduction

Land subsidence caused by the extraction of underground fluid is a problem of major importance to environmental geosciences and environmental geomechanics. The problem has caused damage to infrastructures and resulted in huge economic loss in many regions of the world, posing a threat to sustainable development of cities (Galloway and Burbey 2011; Cui et al. 2016). In China, some coastal soft soil areas, especially the Yangtze River Delta and the North China Plain, have been suffering from severe subsidence due to the fact that groundwater is withdrawn extremely for industrial and domestic purposes (Xue et al. 2005; Chai et al. 2004; Cui and Tang 2010). How to keep the balance between the reasonable use of groundwater resources and the control of land subsidence is a hotspot in geotechnical engineering. Therefore, evaluating and predicting the surface settlement induced by the changes in groundwater levels precisely is the key to this problem.

The extraction of fluid from saturated soil results in decline in pore water pressure and the increase in the effective stress, which induce the deformation of soil skeleton. Some researchers have combined the three-dimensional flow model with one-dimensional consolidation theory to calculate the settlement due to groundwater drawdown, and considered nonlinear seepage and consolidation behaviors of soils (Chen et al. 2003; Shen et al. 2006; Shen and Xu 2011), while this method cannot present the inter-coupling process between soil stress and pore pressure, and it is also hard to illustrate the occurrence of earth fissures, which are caused by horizontal stress and displacement.

However, the general theory of Biot's consolidation, which takes into account the coupling between the soil skeleton and the fluid in a porous medium, can be regarded as the fundamental theory for the analysis of pumping in the saturated soil. Booker and Carter (1987) developed the complete solution for the consolidation of a saturated elastic half-space caused by pumping from a point sink with consideration of

Z.-D. Cui, *Land Subsidence Induced by the Engineering-Environmental Effect*,
https://doi.org/10.1007/978-981-10-8040-1_3

the anisotropy of permeability. Selvadurai and Kim (2015) presented analytical solution for the transient poroelastic problem related to fluid extraction from a circular disk-shaped region located within the half-space, and found that the size of withdrawal region had a direct impact on the surface displacement and pore water pressure. Xie et al. (2014) presumed that the distribution of groundwater level in the aquifer coincided with Theis well model. The excess pore pressure of the surface of this layer was regarded as the boundary condition for the overlying aquitard, and then the influence of the compressibility of the fluid on the consolidation behavior of the aquitard layer was studied. Based on previous researches, we can find that most scholars mainly considered the effects of the compressibility of fluid or soil constituents and the anisotropy of permeability on the consolidation deformation of soil, but actually, the latter acts as a predominant factor in consolidation analysis, while the former has few effects and can even been ignored.

Besides, it is widely known that natural foundation has experienced a sedimentation process and thus has obvious stratification characteristic. But in general, it is acceptable to assume that a single soil layer is homogeneous along the horizontal direction for the convenience of theoretical analysis. Vardoulakis and Harnpatanapanich (1986) and Booker and Small (1987) analyzed the problem of saturated multilayered soil with isotropic permeability caused by a point sink by using a numerical Laplace–Fourier transform inversion technique and the finite layer method, respectively. Chen (2013) utilized integral transform and transfer matrix technique to obtain the steady-state solutions for multilayered and poroelastic half-space subjected to three types of pumping, including a point sink, a circular area, and a ring line. Ai and Zeng (2013) used Laplace–Hankel integral transform and the analytical layer-element method to derive the solutions for saturated multilayered soils caused by a point sink in the transformed domain.

The current researches normally simplify a pumping well into a sink point or a circular extraction region and ignore the length of dewatering well, which is not consistent with practical engineering. Besides, the combined effects of pumping and recharging on the consolidation of stratified soils have not been studied by theoretical methods, but this is extremely important, since it can help us understand deformation behaviors of multilayered soils induced by changes in groundwater levels, and lead us to use groundwater resources reasonably and avoid serious subsidence. Therefore, this paper establishes the analytical model of stratified soils under the effect of pumping from the dewatering well by considering anisotropic permeability and the length of the well. Adopting Laplace–Hankel transforms, the exact expressions of stresses, displacements, excess pore pressure, and seepage velocity can be obtained in the transformed domain. Subsequently, the stiffness matrix of a single soil layer is built to describe the relationship between those components. Then, the global stiffness matrix of layered soils is assembled by using the analytical layer-element method proposed by Ai and Zeng (2013). Finally, the actual solutions can be acquired by the inversion of the Laplace–Hankel transforms. To verify the accuracy of this method, comparisons with the results of ABAQUS are carried out. Moreover, a series of parametric studies, especially the length of the well and the combined effect of pumping and recharging, are conducted to analyze the consolidation behaviors of layered soils with anisotropic permeability (Li and Cui 2017).

3.2 Analytical Solutions for a Single Soil Layer

3.2.1 Governing Equations

As for the axisymmetric Biot's consolidation problem, the force equilibrium equations (with no body forces) expressed in total stresses are

$$\frac{\partial \sigma_{rr}}{\partial r} + \frac{\partial \sigma_{rz}}{\partial z} + \frac{\sigma_{rr} - \sigma_{\theta\theta}}{r} = 0 \tag{3.1a}$$

$$\frac{\partial \sigma_{rz}}{\partial r} + \frac{\partial \sigma_{zz}}{\partial z} + \frac{\sigma_{rz}}{r} = 0, \tag{3.1b}$$

where $\sigma_{rr}, \sigma_{zz}, \sigma_{\theta\theta}$ are total normal components and σ_{rz} is shear stress component.

It is assumed that the solid constituent is incompressible. The constitutive equations take the form

$$\sigma_{ij} = 2G(\varepsilon_{ij} + \frac{v}{1-2v}\varepsilon_v \delta_{ij}) - p\delta_{ij}, \tag{3.2}$$

where the subscripts i and j can be r, z or θ; ε_{ij} are strain components; G and v are shear modulus and Poisson's ratio, respectively; p is excess pore pressure; $\varepsilon_v = \frac{\partial u_r}{\partial r} + \frac{u_r}{r} + \frac{\partial u_z}{\partial z}$ denotes volumetric strain; δ_{ij} is the Kronecker delta.

Substituting Eq. (3.2) into Eqs. (3.1a) and (3.1b), we obtain

$$\left(\nabla^2 - \frac{1}{r^2}\right)u_r + (2\eta - 1)\frac{\partial \varepsilon_v}{\partial r} - \frac{1}{G}\frac{\partial p}{\partial r} = 0 \tag{3.3a}$$

$$\nabla^2 u_z + (2\eta - 1)\frac{\partial \varepsilon_v}{\partial z} - \frac{1}{G}\frac{\partial p}{\partial z} = 0, \tag{3.3b}$$

where $\nabla^2 = \frac{\partial^2}{\partial r^2} + \frac{1}{r}\frac{\partial}{\partial r} + \frac{\partial^2}{\partial z^2}$ is the Laplacian operator; $\eta = \frac{1-v}{1-2v}$.

Assuming that the permeability of porous medium is anisotropic and the fluid constituent is incompressible, the pore fluid mass conservation equation is given by

$$\frac{\partial \varepsilon_v}{\partial t} = \frac{k_h}{\gamma_w}\left(\frac{\partial^2 p}{\partial r^2} + \frac{1}{r}\frac{\partial p}{\partial r}\right) + \frac{k_z}{\gamma_w}\frac{\partial^2 p}{\partial z^2}, \tag{3.4}$$

where k_z and k_h are the vertical and horizontal permeability, respectively; γ_w is the unit weight of water; t is a time variable.

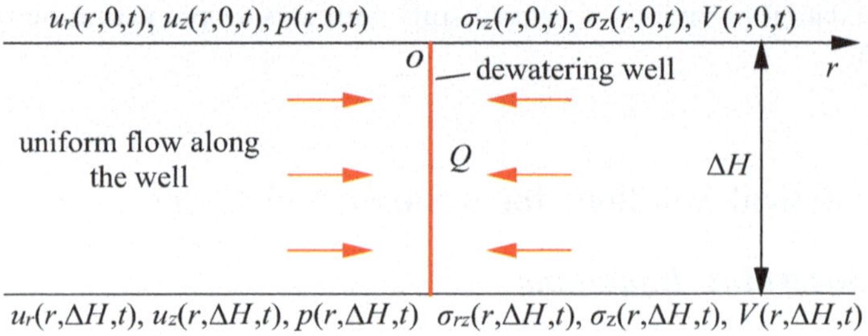

Fig. 3.1 Consolidation of a soil layer due to water extraction from the dewatering well

Figure 3.1 illustrates groundwater drawdown from a dewatering well in a single soil layer. It is assumed that the well is arranged in the whole depth of the soil layer and its radius is infinitely small, and water flows uniformly through the whole well. H represents the vertical coordinate of the undersurface of the soil layer, while ΔH represents its thickness. Therefore, the pumping boundary on the centerline can be expressed as:

$$r\frac{\partial p}{\partial r}\bigg|_{r\to 0}=\frac{Q\gamma_w}{2\pi\Delta Hk_r}, \tag{3.5}$$

where Q is the water flow velocity through the dewatering well and it can be defined as any continuous derivable function, like constants, periodic function, etc. It is worth noting that $Q > 0$ represents water drawdown and $Q < 0$ represents water recharging, while a soil layer does not contain a dewatering well, we can set Q to be 0.

It is assumed that the flow of pore water complies with Darcy's law, so the vertical seepage velocity is

$$V_z=k_z\frac{\partial p}{\partial z} \tag{3.6}$$

3.2.2 Solutions

Applying the operators of $\left(\frac{\partial}{\partial r}+\frac{1}{r}\right)$ and $\frac{\partial}{\partial z}$ to Eqs. (3.3a) and (3.3b), we have

$$\nabla^2\left(\frac{\partial}{\partial r}+\frac{1}{r}\right)u_r+\left(\frac{\partial^2}{\partial r^2}+\frac{1}{r}\frac{\partial}{\partial r}\right)\left((2\eta-1)\varepsilon_v-\frac{p}{G}\right)=0 \tag{3.7a}$$

$$\nabla^2\left(\frac{\partial u_z}{\partial z}\right)+\frac{\partial^2}{\partial z^2}\left((2\eta-1)\varepsilon_v-\frac{p}{G}\right)=0 \tag{3.7b}$$

Combining Eqs. (3.7a) with (3.7b), we have

$$\nabla^2\left(2\eta\varepsilon_v - \frac{p}{G}\right) = 0 \tag{3.8}$$

The Laplace–Hankel transforms are introduced in the partial differential Eqs. (3.2) and (3.4)–(3.8) in order to obtain the ordinary differential equations. The nth-order Laplace–Hankel transform and the corresponding inversion of the function with respect to the variables r and t are defined by Sneddon (1972):

$$\hat{f}_n(\xi, z, s) = \int_0^\infty \int_0^\infty f(r, z, t) r J_n(\xi r)\, \mathrm{e}^{-st} \mathrm{d}r\mathrm{d}t \tag{3.9a}$$

$$f(r, z, t) = \frac{1}{2\pi i} \int_{\gamma - i\infty}^{\gamma + i\infty} \int_0^\infty \hat{f}_n(\xi, z, s)\xi J_n(\xi r)\mathrm{e}^{st}\mathrm{d}\xi\mathrm{d}s \tag{3.9b}$$

where $\hat{f}_n(\xi, z, s)$ is the corresponding variable of $f(r, z, t)$ in the Laplace–Hankel transformed domain; s and ξ are the Laplace and Hankel transform parameters, respectively; $J_n(\xi r)$ denotes the nth-order Bessel function.

Taking Laplace–Hankel transform (zeroth-order) of Eqs. (3.4) and (3.8) and combining the pumping condition Eq. (3.5), we have

$$s\hat{\varepsilon}_{v0} = -\frac{k_h}{\gamma_w}\xi^2 \hat{p}_0 + \frac{k_z}{\gamma_w}\frac{\mathrm{d}^2 p_0}{\mathrm{d}z^2} - \frac{\hat{Q}}{2\pi\Delta H} \tag{3.10}$$

$$\left(\frac{\mathrm{d}^2}{\mathrm{d}z^2} - \xi^2\right)\left(\frac{\mathrm{d}^2\hat{p}_0}{\mathrm{d}z^2} - q^2\hat{p}_0 - \frac{\hat{Q}\gamma_w}{2\pi\Delta H k_z}\right) = 0, \tag{3.11}$$

where $\hat{Q} = \frac{Q}{s}$ denotes the Laplace transform of water flow velocity Q; $q^2 = \alpha\xi^2 + \frac{\gamma_w s}{Mk_z}$, in which $\alpha = \frac{k_h}{k_z}$ and $M{=}2\eta G$.

The solutions to Eqs. (3.10) and (3.11) are

$$\hat{p}_0 = A_1 e^{-\xi z} + A_2 e^{\xi z} + A_3 e^{-qz} + A_4 e^{qz} - Q_1(\xi, \Delta H, s) \tag{3.12}$$

$$\hat{\varepsilon}_{v0} = A_1\varphi e^{-\xi z} + A_2\varphi e^{\xi z} + A_3\frac{e^{-qz}}{M} + A_4\frac{e^{qz}}{M} - \frac{\gamma_w}{Mk_h\xi^2 + \gamma_w s}\frac{\hat{Q}}{2\pi\Delta H}, \tag{3.13}$$

where $A_1 \sim A_6$ are arbitrary functions determined by the boundary conditions and continuity condition of adjacent soil layers; $Q_1(\xi, \Delta H, s) = \frac{M}{Mk_h\xi^2 + \gamma_w s}\frac{\hat{Q}\gamma_w}{2\pi\Delta H}$; $\varphi = \frac{\xi^2(1-\alpha)k_z}{s\gamma_w}$.

Taking the first-order, zeroth-order, and zeroth-order Laplace–Hankel transform of Eqs. (3.3a) and (3.3b) and volumetric strain function ε_v, respectively, we have

$$\left(\frac{\mathrm{d}^2}{\mathrm{d}z^2}-\xi^2\right)\hat{u}_{r1}-\xi\left((2\eta-1)\hat{\varepsilon}_{v0}-\frac{1}{G}\hat{p}_0\right)=0 \tag{3.14a}$$

$$\left(\frac{\mathrm{d}^2}{\mathrm{d}z^2}-\xi^2\right)\hat{u}_{z0}+\frac{\mathrm{d}}{\mathrm{d}z}\left((2\eta-1)\hat{\varepsilon}_{v0}-\frac{1}{G}\hat{p}_0\right)=0 \tag{3.14b}$$

$$\hat{\varepsilon}_{v0}=\xi\hat{u}_{r1}+\frac{\mathrm{d}\hat{u}_{z0}}{\mathrm{d}z} \tag{3.14c}$$

Substituting Eqs. (3.12) and (3.13) into Eqs. (3.14a), (3.14b) and (3.14c), the solutions to $\hat{u}_{r1}$ and $\hat{u}_{z0}$ in the transformed domain can be expressed as

$$\begin{aligned}\hat{u}_{r1}=&\frac{1-(M-G)\varphi}{2G}ze^{-\xi z}A_1-\frac{1-(M-G)\varphi}{2G}ze^{\xi z}A_2+\frac{\xi e^{-qz}}{M(-q^2+\xi^2)}A_3\\&+\frac{\xi e^{qz}}{M(-q^2+\xi^2)}A_4+e^{-\xi z}A_5+e^{\xi z}A_6-Q_2(\xi,\Delta H,s)\end{aligned} \tag{3.15a}$$

$$\begin{aligned}\hat{u}_{z0}=&\frac{1-G\varphi-M\varphi+z\xi(1+G\varphi-M\varphi)}{2G\xi}e^{-\xi z}A_1+\frac{G\varphi+M\varphi-1+z\xi(1+G\varphi-M\varphi)}{2G\xi}e^{\xi z}A_2\\&-\frac{qe^{-qz}}{M(q^2-\xi^2)}A_3+\frac{qe^{qz}}{M(q^2-\xi^2)}A_4+e^{-\xi z}A_5-e^{\xi z}A_6,\end{aligned} \tag{3.15b}$$

where $Q_2(\xi,\Delta H,s)=\frac{1}{Mk_h\xi^2+\gamma_w s}\frac{\hat{Q}\gamma_w}{2\pi\Delta H\xi}$.

From Eq. (3.2), we have

$$\sigma_{rz}=G\left(\frac{\partial u_r}{\partial z}+\frac{\partial u_z}{\partial r}\right) \tag{3.16a}$$

$$\sigma_z=2G\left(\frac{\partial u_z}{\partial z}+(\eta-1)\varepsilon_v\right)-p_0 \tag{3.16b}$$

Similarly, taking the first-order, zeroth-order, and zeroth-order Laplace–Hankel transform of Eqs. (3.16a), (3.16b) and (3.6), respectively, we have

$$\hat{\sigma}_{rz1}=G\left(\frac{\mathrm{d}\hat{u}_{r1}}{\mathrm{d}z}-\xi\hat{u}_{z0}\right) \tag{3.17a}$$

$$\hat{\sigma}_{z0} = 2G\left(\frac{\mathrm{d}\hat{u}_{z0}}{\mathrm{d}z} + (\eta - 1)\hat{\varepsilon}_{v0}\right) - \hat{p}_0 \tag{3.17b}$$

$$\hat{V}_0 = -\xi k_z e^{-\xi z} + \xi k_z e^{\xi z} - q k_z e^{-qz} + q k_z A_4 e^{qz} \tag{3.17c}$$

Substituting Eqs. (3.12), (3.15a) and (3.15b) into Eqs. (3.17a) and (3.17b), we have

$$\begin{aligned}\hat{\sigma}_{rz1} = {} & (G\varphi - z\xi(1 + G\varphi - M\varphi))e^{-\xi z}A_1 - (G\varphi + z\xi(1 + G\varphi - M\varphi))e^{\xi z}A_2 \\ & + \frac{2q\xi G e^{-qz}}{M(q^2 - \xi^2)}A_3 - \frac{2q\xi G e^{qz}}{M(q^2 - \xi^2)}A_4 - 2e^{-\xi z}G\xi A_5 + 2e^{\xi z}G\xi A_6\end{aligned} \tag{3.18a}$$

$$\begin{aligned}\hat{\sigma}_{z0} = {} & (-1 + M\varphi - z\xi(1 + G\varphi - M\varphi))e^{-\xi z}A_1 - (1 - M\varphi - z\xi(1 + G\varphi - M\varphi))e^{\xi z}A_2 \\ & + \frac{2\xi^2 G e^{-qz}}{M(q^2 - \xi^2)}A_3 + \frac{2\xi^2 G e^{qz}}{M(q^2 - \xi^2)}A_4 - 2e^{-\xi z}G\xi A_5 - 2e^{\xi z}G\xi A_6 + Q_3(\xi, \Delta H, s),\end{aligned} \tag{3.18b}$$

where $Q_3(\xi, \Delta H, s) = \frac{2G}{Mk_h\xi^2 + \gamma_w s}\frac{\hat{Q}\gamma_w}{2\pi\Delta H}$.

In order to establish relationships between stresses, displacements, excess pore pressure, and seepage velocity in the transformed domain, we can define

$$\hat{\mathrm{U}}(\xi, z, s) = [\hat{u}_{r1}(\xi, z, s), \hat{\sigma}_{z0}(\xi, z, s), \hat{p}_0(\xi, z, s)]^{\mathrm{T}} \tag{3.19a}$$

$$\hat{\Gamma}(\xi, z, s) = \left[\hat{\sigma}_{rz1}(\xi, z, s), \hat{u}_{z0}(\xi, z, s), \hat{V}_0(\xi, z, s)\right]^{\mathrm{T}} \tag{3.19b}$$

$$T(\xi, \Delta H, s) = [Q_2(\xi, \Delta H, s), -Q_3(\xi, \Delta H, s), Q_1(\xi, \Delta H, s)]^{\mathrm{T}} \tag{3.19c}$$

If a soil layer does not contain a dewatering well, $T(\xi, \Delta H, s) = [0,\ 0,\ 0]^{\mathrm{T}}$.

Obviously, $\hat{\mathrm{U}}(\xi, 0, s) + T(\xi, \Delta H, s)$, $\hat{\Gamma}(\xi, 0, s)$, $\hat{\mathrm{U}}(\xi, H, s) + T(\xi, \Delta H, s)$ and $\hat{\Gamma}(\xi, H, s)$ are all vectors made of $A_1 \sim A_6$, so the relationship can be expressed as

$$\begin{bmatrix} -\hat{U}(\xi, 0, s) \\ \hat{U}(\xi, H, s) \end{bmatrix} + \begin{bmatrix} -T(\xi, \Delta H, s) \\ T(\xi, \Delta H, s) \end{bmatrix} = R \cdot [A_1, A_2, A_3, A_4, A_5, A_6]^{\mathrm{T}} \tag{3.20a}$$

$$\begin{bmatrix} \hat{\Gamma}(\xi, 0, s) \\ \hat{\Gamma}(\xi, H, s) \end{bmatrix} = S \cdot [A_1, A_2, A_3, A_4, A_5, A_6]^{\mathrm{T}} \tag{3.20b}$$

Combining Eqs. (3.20a) with (3.20b), we have

$$\begin{bmatrix} -\hat{\mathrm{U}}(\xi,0,s) \\ \hat{\mathrm{U}}(\xi,H,s) \end{bmatrix} + \begin{bmatrix} -T(\xi,\Delta H,s) \\ T(\xi,\Delta H,s) \end{bmatrix} = \Phi \cdot \begin{bmatrix} \hat{\Gamma}(\xi,0,s) \\ \hat{\Gamma}(\xi,H,s) \end{bmatrix} \tag{3.21}$$

where $\Phi = R \cdot S^{-1}$ is a stiffness matrix of order 6×6, which describes the relationship between stresses, displacements, excess pore pressure, and seepage velocity of the top and the bottom surface of a soil layer in the transformed domain. The expression of Φ can be found in Appendix A.

3.3 Numerical Assembly of Multilayered Soils

A stratified soil system, as shown in Fig. 3.2, consists of n layers, so the number of unknowns is $6n+6$. The thickness of the ith layer is $\Delta H_i = H_i - H_{i-1}(H_0 = 0)$, in which H_i is the vertical coordinate of the bottom surface of the ith layer. It is worth noting that if the dewatering well is not arranged in the whole depth of one soil layer, we can partition this layer into several new sublayers along the horizontal direction. One of them is a layer that the well distributes in the whole depth of, and the others do not contain the well.

It is assumed that the surface of the top layer is permeable and stress-free, and the bottom surface of the last layer is impermeable and fixed, so the boundary conditions in the transformed domain are

$$\hat{\sigma}_{z0}(\xi,0,s) = \hat{\sigma}_{rz1}(\xi,0,s) = p_0(\xi,0,s) = 0 \tag{3.22a}$$

$$\hat{u}_{r1}(\xi,H_n,s) = \hat{u}_{z0}(\xi,H_n,s) = \hat{V}_0(\xi,H_n,s) = 0 \tag{3.22b}$$

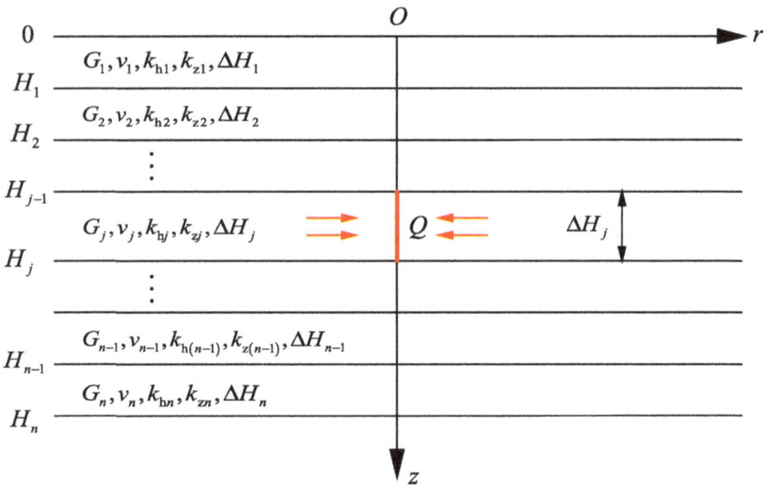

Fig. 3.2 Multilayered soils containing a dewatering well

Besides, the interfaces of the two adjacent layers satisfy the continuity conditions, so we have

(1) the *j*th soil layer including a dewatering well

$$\hat{\Gamma}\left(\xi, H_{j-1}^{-}, s\right) = \hat{\Gamma}\left(\xi, H_{j-1}^{+}, s\right),\ \hat{\mathrm{U}}\left(\xi, H_{j-1}^{-}, s\right) = \hat{\mathrm{U}}\left(\xi, H_{j-1}^{+}, s\right) + T(\xi, \Delta H_j, s) \tag{3.23a}$$

$$\hat{\Gamma}\left(\xi, H_j^{+}, s\right) = \hat{\Gamma}\left(\xi, H_j^{-}, s\right), \hat{\mathrm{U}}\left(\xi, H_j^{+}, s\right) = \hat{\mathrm{U}}\left(\xi, H_j^{-}, s\right) + T(\xi, \Delta H_j, s) \tag{3.23b}$$

(2) the other soil layers

$$\hat{\Gamma}\left(\xi, H_{j-1}^{-}, s\right) = \hat{\Gamma}\left(\xi, H_{j-1}^{+}, s\right),\ \hat{\mathrm{U}}\left(\xi, H_{j-1}^{-}, s\right) = \hat{\mathrm{U}}\left(\xi, H_{j-1}^{+}, s\right) \tag{3.24a}$$

$$\hat{\Gamma}\left(\xi, H_j^{+}, s\right) = \hat{\Gamma}\left(\xi, H_j^{-}, s\right), \hat{\mathrm{U}}\left(\xi, H_j^{+}, s\right) = \hat{\mathrm{U}}\left(\xi, H_j^{-}, s\right) \tag{3.24b}$$

where H_j^- and H_j^+ denote the depth H_j of the *j*th layer and the (*j* + 1)th layer, respectively.

Applying Eq. (3.21) to each layer and on the superposition of each layer stiffness matrix, the global stiffness matrix of the multilayered soils is assembled by taking into account the continuity condition:

$$\begin{bmatrix} -\hat{\mathrm{U}}(\xi,0,s) \\ 0 \\ \vdots \\ T(\xi,\Delta H_j,s) \\ -T(\xi,\Delta H_j,s) \\ \vdots \\ 0 \\ \hat{\mathrm{U}}(\xi,H_n,s) \end{bmatrix} = \begin{bmatrix} \Phi^{(1)} & & & & 0 \\ & \Phi^{(2)} & & & \\ & & \ddots & & \\ & & & \Phi^{(n-1)} & \\ \text{Symmetric} & & & & \Phi^{(n)} \end{bmatrix}_{(3n+3)\times(3n+3)} \begin{bmatrix} \hat{\Gamma}(\xi,0,s) \\ \hat{\Gamma}(\xi,H_1,s) \\ \vdots \\ \hat{\Gamma}(\xi,H_{j-1},s) \\ \hat{\Gamma}(\xi,H_j,s) \\ \vdots \\ \hat{\Gamma}(\xi,H_{n-1},s) \\ \hat{\Gamma}(\xi,H_n,s) \end{bmatrix} \tag{3.25}$$

By combining 6 boundary conditions, namely, Eqs. (3.22a) and (3.22b), we can derive $3n+3$ unknowns $\left[\hat{\Gamma}(\xi,0,s)\ldots\hat{\Gamma}(\xi,H_j,s))\ldots\hat{\Gamma}(\xi,H_n,s)\right]$ through solving Eq. (3.25), so the left $3n-3$ unknowns $\left[\hat{\mathrm{U}}(\xi,H_1,s)\ \ldots\ \hat{\mathrm{U}}(\xi,H_j,s)\ \ldots\ \hat{\mathrm{U}}(\xi,H_{n-1},s)\right]$

can be obtained by substituting $3n+9$ known quantities into Eq. (3.21). Now, the solutions of stresses, displacements, excess pore pressure, and seepage velocity at any point of any interface of layered soil system are available in the transformed domain. In order to obtain the real solutions, the Talbot method (Talbot 1979) and the technique proposed by Ai et al. (2002) are employed for the inversion of the Laplace transforms and the Hankel transforms, respectively.

3.4 Verification of Present Method

To verify the feasibility and accuracy of the proposed analytical solutions, the results of theoretical calculation are compared with those obtained from the finite-element software ABAQUS in terms of the surface displacements and excess pore pressure on the horizontal plane $z = \Delta H_1$. The axisymmetric two-layered soil model including a dewatering well arranged in the whole depth of the aquifer is shown in Fig. 3.3a, where soil parameters are also indicated. In the numerical model, we suppose that the horizontal dimension of layered soils is large enough to simulate the infinite boundary. Besides, the radius of the dewatering well is 0.1 m, which can be nearly ignored compared with other geometrical dimensions, and the constant water flow velocity through the well is regarded as the pumping condition. The elastic modulus of soils can be obtained according to $E = 2G(1+v)$ and the void ratio is 0.6. The other assumptions are as same as those of the theoretical model. For the convenience, we define the following dimensionless quantities, which are also valid in the next section:

$$\bar{u}_z = \frac{u_z k_{z1} \eta_1 G_1}{\gamma_w Q}, \bar{u}_r = \frac{u_r k_{z1} \eta_1 G_1}{\gamma_w Q}, \bar{p}_0 = \frac{p_0 k_{z1} \Delta H_1}{\gamma_w Q}, \bar{t} = \frac{t \eta_1 G_1 k_{z1}}{\gamma_w \Delta H_1^2}, \bar{r} = \frac{r}{\Delta H_1}, \bar{z} = \frac{z}{\Delta H_1}.$$

In Fig. 3.3a, b, c, we can find that the present method agrees well with the numerical simulation, no matter at the early stage of consolidation or at the end. Apart from this, the largest vertical displacement and excess pore pressure occur in the centerline, while the figure for horizontal displacement exists in the range of $\bar{r}$ between 1 and 2 and its position moves outward over time.

3.5 Parametric Studies and Numerical Results

3.5.1 The Length of the Dewatering Well

Selvadurai and Kim (2015) investigated the effects of the radius of the dewatering well (ignoring the well length) on the surface settlement and dissipation of excess

Fig. 3.3 Comparison between theoretical calculations and the numerical results of ABAQUS

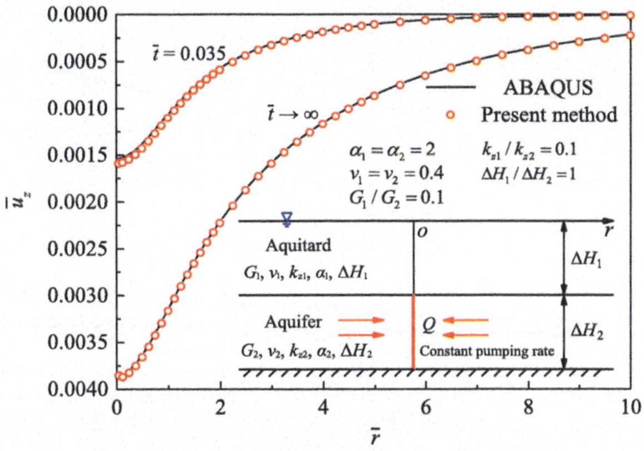

(a) vertical displacement

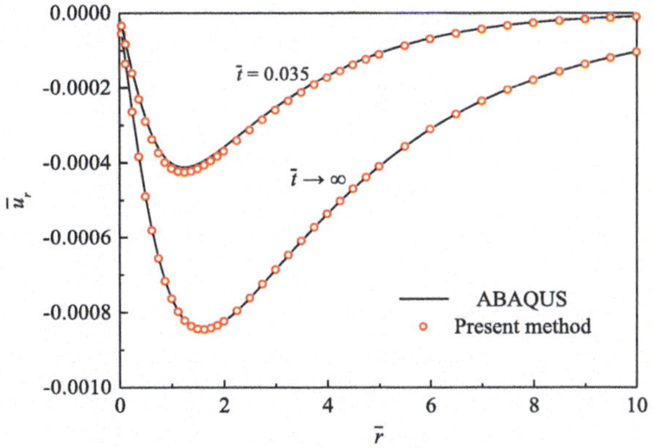

(b) horizontal displacement

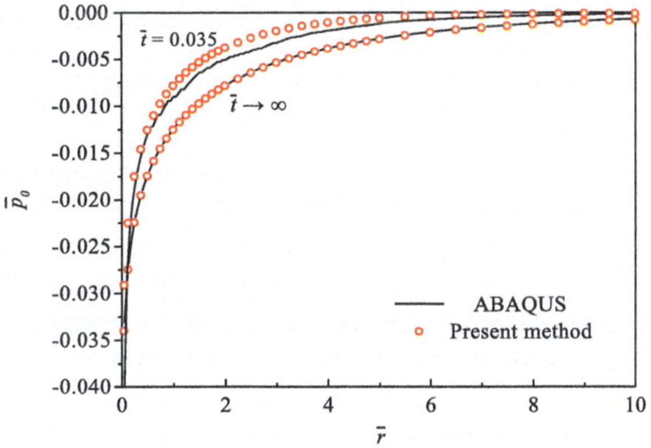

(c) excess pore water pressure

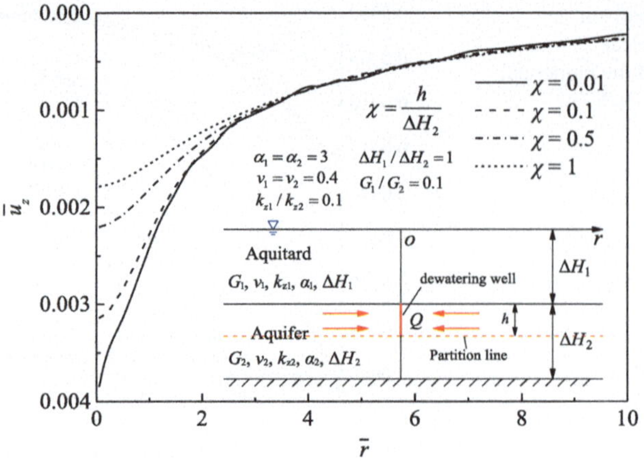

(a) vertical displacement

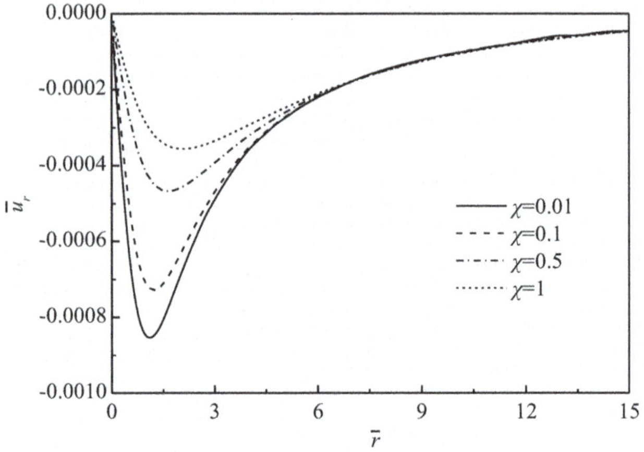

(b) horizontal displacement

Fig. 3.4 The influence of the length of a dewatering well on the surface displacements

pore pressure. In this chapter, we focus on the length of the well h and define it in a dimensionless form $\chi = h/\Delta H_1$, where ΔH_1 is the thickness of the aquitard. In Fig. 3.4, the vertical and horizontal surface displacements are plotted against dimensionless radial distance. Because the length of dewatering well is not equal to the thickness of the aquifer, we should partition this layer for obtaining a new sublayer with the well arranged along the whole thickness. When χ is 0.01, we can appropriately regard the dewatering well as a point sink, and the vertical and horizontal displacement at this level of χ is significantly higher than others, nearly more than twice the figures for $\chi = 1$. The increase in χ induces the smaller surface displacements. This means that the length of dewatering well has an essential impact on the deformation behaviors of multilayered soils caused by groundwater drawdown and deserves our considerations in this problem, while regarding the well as a point sink normally leads to the relatively large calculated results.

3.5.2 Anisotropic Permeability

Anisotropic permeability is a basic characteristic of soils and it has a bigger influence on the consolidation analysis than other factors do, like the compressibility of fluid or soil constituents (Xie et al. 2014) as well as elastic anisotropic (Tarn and Lu 1991). Three values of anisotropic permeability parameter $\alpha = 1, 3, 10$ are selected to study their influence on the surface displacements and the excess pore pressure on the horizontal plane $z = \Delta H_1$. The soil system contains three layers and the relevant parameter relationships as shown in Fig. 3.5. It can be seen that the increase in α induces significantly smaller surface displacements and excess pore pressure, since the larger horizontal permeability facilitates the main supply of pore water from the horizontal direction instead of water seepage in the vertical direction.

3.5.3 Layered Characteristic

Figure 3.6 presents the consolidation of multilayered soils caused by pumping and is compared with that of a single soil layer with weighted parameters at different times. At the initial stage of consolidation, there is few difference between the uniform soil layer and the stratified soil in terms of surface displacements and excess pore pressure on the horizontal plane $z = \Delta H_1$. But this difference widens obviously over time. Especially when the consolidation is finished, the vertical and horizontal displacements calculated by the present method were more than twice of those obtained by the weighted method. Therefore, we cannot transform the layered soil into the uniform soil by taking a weighted average of parameters to calculate the final settlement and the excess pore pressure, which may be quite different from the real values.

Another aspect that can reflect the layered characteristic of soils is that groundwater is extracted from different aquifers. As shown in Fig. 3.7, a four-layered soil system contains two aquifers and all the parameters of these two layers are the same, so we can study the influence of the depth of pumping layer on the surface displacements. For the vertical surface displacement, pumping from the shallow aquifer induces the larger settlement than that from the deep aquifer in the range of $\bar{r}$ between 0 and 2.5, in other words, the former can result in more serious differential settlement in the surrounding areas of centerline. However, the latter can lead to the larger influence range of settlement. There is a similar trend seen in the horizontal surface displacement. Along with a shift of pumping from the shallow

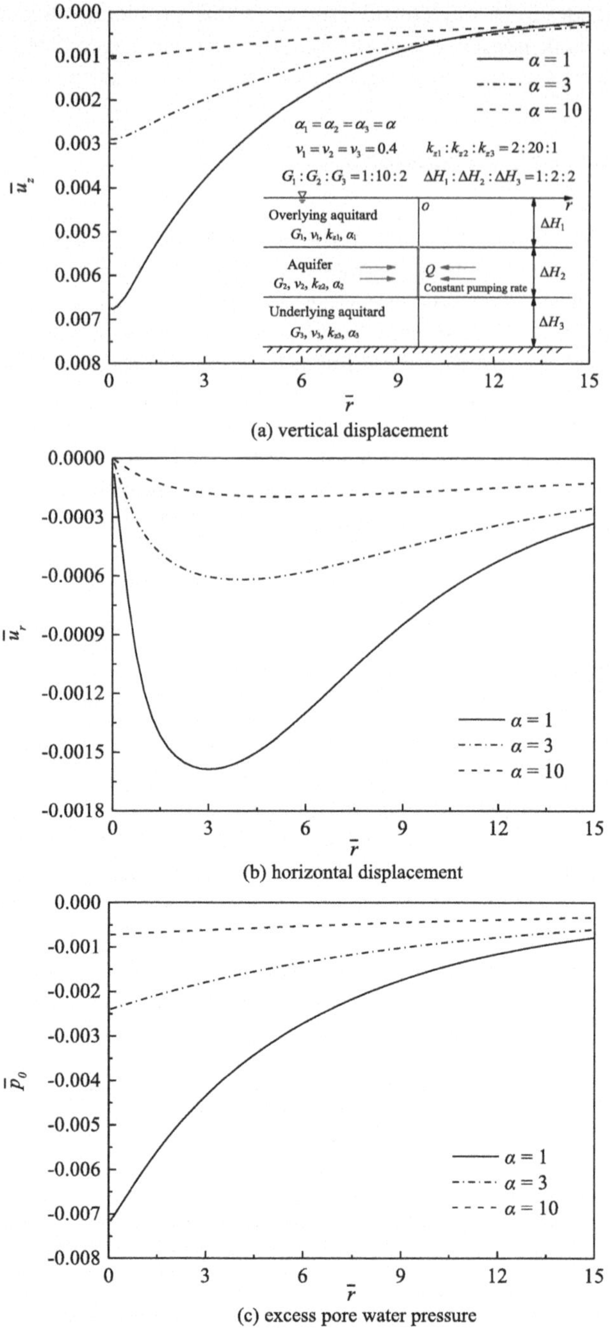

Fig. 3.5 The influence of anisotropic permeability on the surface displacements and the excess pore pressure of the top surface of the aquifer

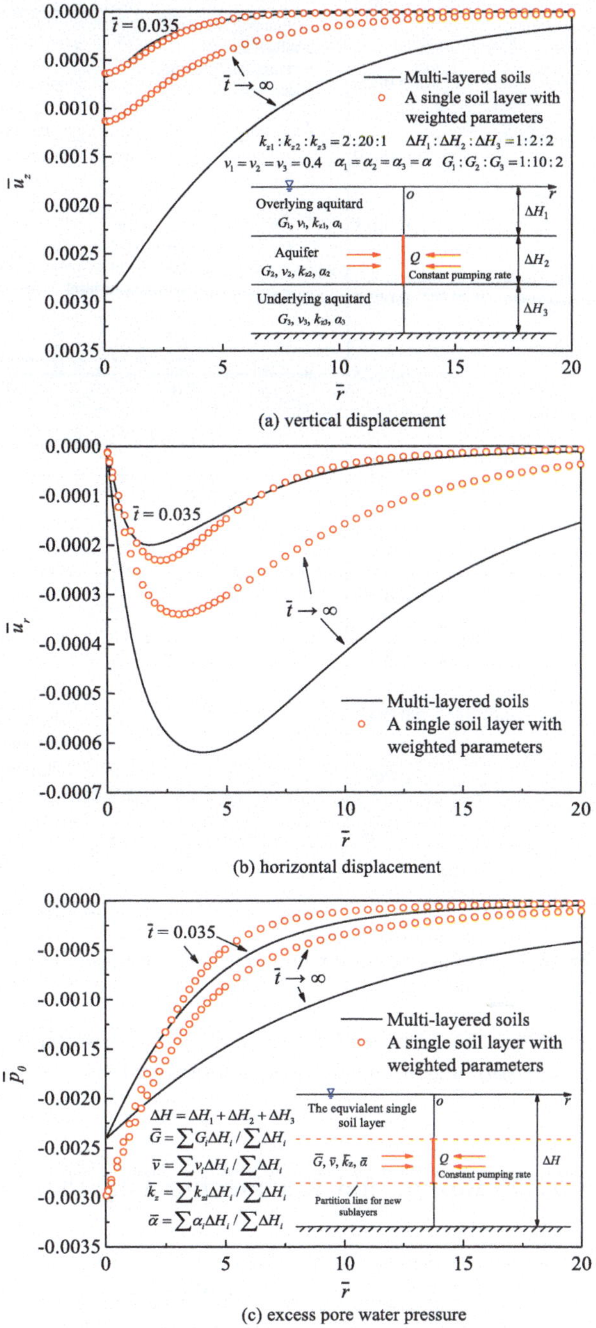

Fig. 3.6 Comparison between multilayered soils and a single-layer soil with weighted parameters

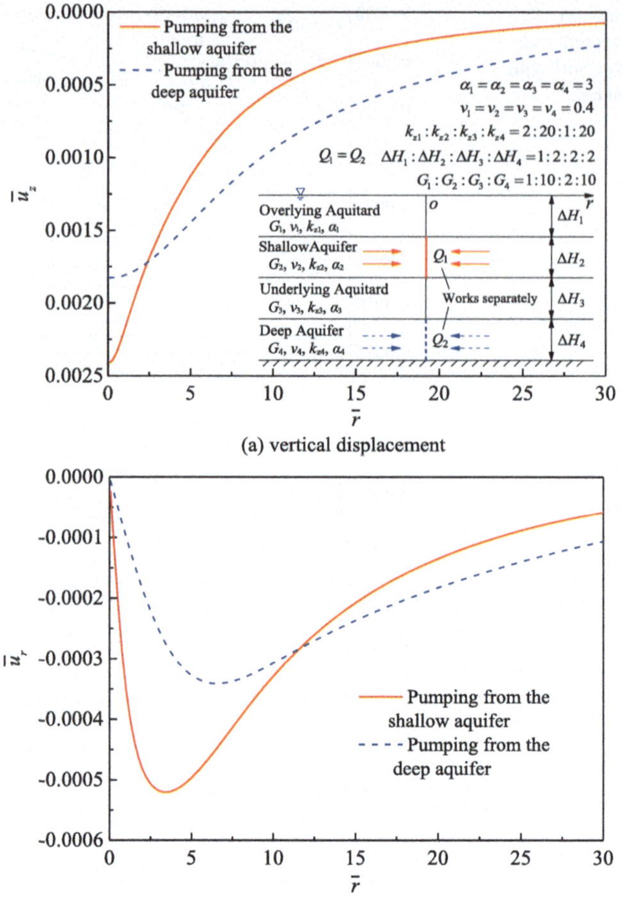

(a) vertical displacement

(b) horizontal displacement

Fig. 3.7 The influence of different layers of dewatering on the surface displacements

aquifer to the deep aquifer, the maximum horizontal surface displacement decreases remarkably and its position moves outward.

3.6 Pumping and Recharging Together

Previous researches mainly focus on the influence of the single factor such as pumping or recharging on the consolidation problems, but their combined effect has been rarely considered. In this section, two situations that pumping and recharging work together are selected to understand this effect. One situation is that water is extracted from the deep aquifer and recharged into the shallow aquifer at the same time (Fig. 3.8), the other is that the water flow velocity on the surface of dewatering well is a periodical function, which means that pumping and recharging work alternatively, as shown in Fig. 3.9.

For the first situation, the recharging ratio θ is defined as $|Q_1|/|Q_2|$ to describe the relationship between recharging and pumping, where $|Q_1|$ and $|Q_2|$ denote the recharging amount and the pumping amount, respectively. $\theta = 0$ means there is no recharging, while $\theta = 1$ means the amounts of pumping and recharging reach the balance. Four values of $\theta = 0, 0.1, 0.3, 1$ are selected to study their influence on the surface displacements. In Fig. 3.8, it can be observed that the increase in the recharging ratio θ enables vertical displacement and horizontal displacement to decrease dramatically. When it reaches the balance between pumping and recharging, the surface rebound and the outward horizontal surface movement

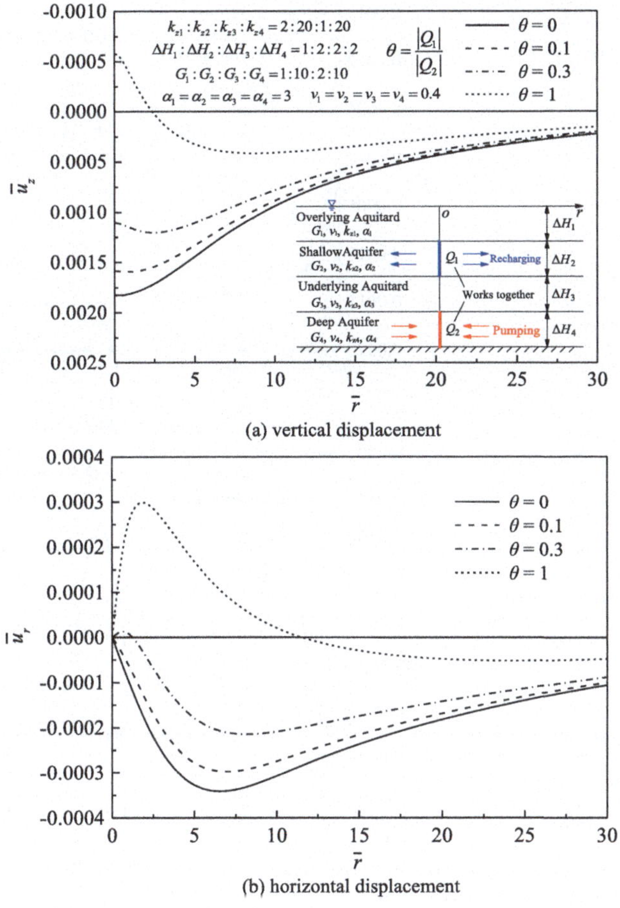

Fig. 3.8 The influence of the amount of water recharging on the surface displacements

Fig. 3.9 The influence of periodical water pumping rate on the surface displacements at the point $\bar{r} = 0.5$ and excess pore pressure at the point $\bar{r} = 0.5$ of the horizontal plane $z = \Delta H_1$

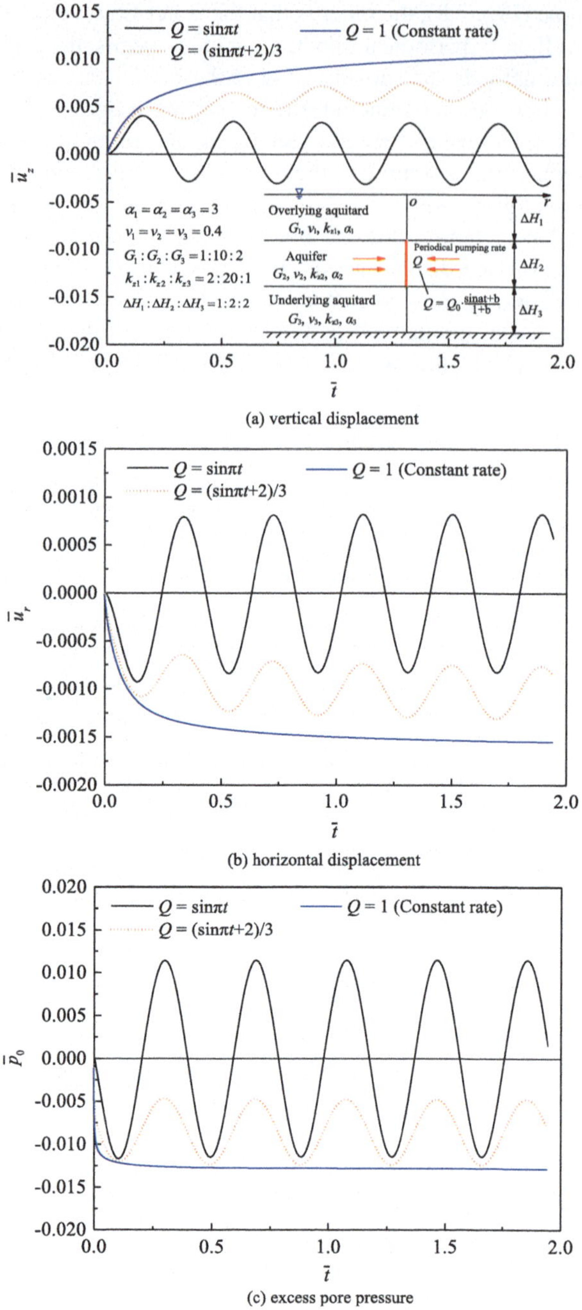

occur in the surrounding area of centerline. Therefore, artificially recharging is an effective method to control land subsidence and the recharging ratio θ is the key to this problem.

In Shanghai, the method of pumping in summer and recharging in winter has been conducted to control subsidence (Gong 2009), so it is necessary to figure out the deformation behaviors of multilayered soils caused by cyclic pumping and recharging through theoretical analysis. The periodical water pumping function Q is defined as $Q = Q_0(\sin \mathrm{a}t + \mathrm{b})/(1 + \mathrm{b})$, where Q_0 denotes the pumping amplitude, a is the pumping period and b is the pumping deviator. We select three types of pumping functions, namely, $Q = 1$, $Q = \sin \pi t$, and $Q = (\sin \pi t + 2)/3$, which represent the constant pumping rate, the periodical pumping rate, and the periodical pumping rate with a pumping deviator, respectively. In Fig. 3.9a, b, it can be seen that the changes of vertical displacement and horizontal displacement with time induced by the pumping function $Q = \sin \pi t$ also comply with the trigonometric function but their amplitude and period do not change over time, while the displacements caused by the constant pumping function $Q = 1$ accumulate gradually. The surface displacements induced by the pumping function $Q = (\sin \pi t + 2)/3$ contain all the characteristics of two previous functions, namely, the periodical fluctuation and the accumulated deformation. Besides, the displacements induced by the periodical pumping rate are smaller than that caused by the constant pumping rate, although their pumping amplitudes are the same. Combined with Fig. 3.9c, we can find that even if the excess pore pressure caused by $Q = 1$ remains stable after the time $\bar{t} = 1$, the displacements still develop further. There is a similar trend seen in the function $Q = (\sin \pi t + 2)/3$. This means that the change of displacements induced by pumping lags behind that of the excess pore pressure.

3.7 Conclusions

The chapter presents transient solutions to the consolidation problems of multilayered soils with anisotropic permeability caused by groundwater drawdown and recharging. Based on the Laplace–Hankel transforms, we obtain the exact expression of stresses, displacements, excess pore pressure, and vertical seepage velocity in the transformed domain. Subsequently, the stiffness matrix of a single layer is established to describe the relationships between these six components. Then, combined with the continuity conditions of adjacent layers and the boundary conditions, the global stiffness matrix equation of layered soils is assembled. At last, the real solutions can be obtained by the inversion of Laplace–Hankel transforms. The theoretical calculation results are in good agreement with numerical results of ABAQUS, which proves the validity of present method. Some numerical examples are provided to investigate the influence of the changes of parameters on the consolidation behaviors of stratified soils, and also compared with existing research achievements. The results include the following:

(1) The length of dewatering well should be considered in the consolidation analysis of layered soils caused by water pumping from the well. The increase in the length of the well leads to smaller surface displacement.
(2) The larger values of anisotropic permeability parameters α can lead to the smaller surface displacements and excess pore pressure.
(3) The layered soil cannot be regarded easily as the uniform soil by taking a weighted average of soil parameters, since this method leads to great deviation of the description of consolidation behaviors of layered soils.
(4) Pumping from the shallow aquifer can result in the more serious differential settlement in the surrounding areas of centerline, while pumping from the deep aquifer can lead to the larger influence range of settlement.
(5) The increase in the recharging ratio θ can effectively restrict the vertical surface settlement and horizontal surface displacement. This effect is particularly obvious when the amounts of pumping and recharging reach the balance.
(6) The displacements induced by the cyclic pumping and recharging are smaller than those caused by pumping alone, although their pumping amplitudes are the same.
(7) The change of surface displacements induced by pumping lags behind that of the excess pore pressure.

References

Ai ZY, Zeng WZ (2013) Consolidation analysis of saturated multi-layered soils with anisotropic permeability caused by a point sink. Int J Numer Anal Met 37(7):758–770

Ai ZY, Yue ZQ, Tham LG, Yang M (2002) Extended Sneddon and Muki solutions for multilayered elastic materials. Int J Eng Sci 40(13):1453–1483

Booker JR, Carter JP (1987) Elastic consolidation around a point sink embedded in a half-space with anisotropic permeability. Int J Numer Anal Met 11(1):61–77

Booker JR, Small JC (1987) A method of computing the consolidation behavior of layered soils using direct numerical inversion of Laplace transforms. Int J Numer Anal Met 11(4):363–380

Chai JC, Shen SL, Zhu HH, Zhang XL (2004) Land subsidence due to groundwater drawdown in Shanghai. Geotechnique 54(2):143–147

Chen GJ (2013) Analysis of pumping in multilayered and poroelastic half space. Comput Geotech 30(1):1–26

Chen C, Pei S, Jiao J (2003) Land subsidence caused by groundwater exploitation in Suzhou City, China. Hydrogeol J 11(2):275–287

Cui ZD, Tang YQ (2010) Land subsidence and pore structure of soils caused by the high-rise building group through centrifuge model test. Environ Geol 113(1):44–52

Cui ZD, Li Z, Jia YJ (2016) Model test study on the subsidence of high-rise building group due to variation of groundwater level. Nat Hazards 84(1):35–53

Galloway DL, Burbey TJ (2011) Review: regional land subsidence accompanying groundwater extraction. Hydrogeol J 19(8):1459–1486

Gong SL (2009) Change of groundwater seepage field and its influence on development of land subsidence in Shanghai (in chinese). J Water Resour Water Eng 20(3):1–6

Li Z, Cui ZD (2017) Axisymmetric consolidation of saturated multi-layered soils with anisotropic permeability due to well pumping. Comput Geotech 92:229–239

Selvadurai APS, Kim J (2015) Ground subsidence due to uniform fluid extraction over a circular region within an aquifer. Adv Water Resour 78:50–59

Shen SL, Xu YS (2011) Numerical evaluation of land subsidence induced by groundwater pumping in Shanghai. Can Geotech J 48(9):1378–1392

Shen SL, Xu YS, Hong ZS (2006) Estimation of land subsidence based on groundwater flow model. Mar Georesour and Geotec 24(2):149–167

Sneddon IN (1972) The use of integral transforms. McGraw-Hill, New York

Talbot A (1979) The accurate numerical inversion of Laplace transforms. IMA J Appl Math 23 (1):97–120

Tarn JQ, Lu CC (1991) Analysis of subsidence due to a point sink in an anisotropic porous elastic half space. Int J Numer Anal Met 15(8):573–592

Vardoulakis I, Harnpattanapanich T (1986) Numerical Laplace-Fourier transform inversion technique for layered-soil consolidation problems: I. Fundamental solutions and validation. Int J Numer Anal Met 10(4):347–365

Xie KH, Huang DZ, Wang YL, Deng YB (2014) Axisymmetric consolidation of a poroelastic soil layer with a compressible fluid constituent due to groundwater drawdown. Comput Geotech 56:11–15

Xue YQ, Zhang Y, Ye SJ, Wu JC, Li QF (2005) Land subsidence in China. Environ Geol 48 (6):713–720

Chapter 4
In-Site Monitoring of Land Subsidence

4.1 Introduction

The in-site monitoring data of land subsidence in Shanghai showed that the deformation of the aquifer occupied a large proportion of the total subsidence and it was not synchronized with the changing of the groundwater table (Zhang et al. 2003). In the early period of land subsidence in Shanghai, there was a strong correlation between the land subsidence and the groundwater exploitation. With the construction of modern cities, the correlation between them was weakened gradually. The land subsidence was accelerated because a large number of municipal works and high-rise buildings were constructed in Shanghai. Some researchers proposed that large-scale city constructions were a significant factor for the land subsidence and this factor accounted for about 30% of the total land subsidence in Shanghai (Gong 1998).

Land subsidence caused by building loads was studied and the additional stress caused by building loads in the foundation was the main reason for changing the stress state of the shallow soil (Tang et al. 2007; Jie et al. 2007). The land subsidence caused by building loads was mainly concentrated in the single building as well as the dense high-rise building group, which was studied by the theoretical calculation, the in-site monitoring, the laboratory tests, and the numerical simulation.

In recent years, land subsidence has greatly accelerated though the exploitation of groundwater is strictly controlled in Shanghai. The mechanism of land subsidence in the soft soil areas caused by the dense high-rise buildings was studied by the in situ monitoring, the general model test, and the centrifugal model test (Tang et al. 2008; Cui and Tang 2011), which gave a reference to study the mechanism of land subsidence in the soft soil areas caused by the dense high-rise building group.

In this chapter, the land subsidence caused by four high-rise buildings including World Financial Center (SWFC), Bank of China Tower (BOC Tower), Shanghai

Z.-D. Cui, *Land Subsidence Induced by the Engineering-Environmental Effect*,
https://doi.org/10.1007/978-981-10-8040-1_4

Merchants Tower, and Jinmao Tower, located in Lujiazui area of Shanghai, was studied by the in-site monitoring data. In addition, the relationship between the changing of the groundwater table and the deformation characteristics of soils was analyzed. The numerical simulation was conducted to study the land subsidence caused by two high-rise buildings with relatively close distance, Jinmao Tower and World Financial Center.

4.2 In Situ Monitoring

4.2.1 Geological and Hydrogeological Background

Shanghai is situated at the fore of the Yangtze River delta. The Quaternary deposit is thick and generally reaches 250–300 m in the urban area. The upper layer about 0–150 m is mainly composed of soft-plastic clay and sandy soil. Figure 4.1 shows the drill column section of soil layers in Lujiazui area of Shanghai.

In order to better monitor the land subsidence in Shanghai, a large number of monitoring system have been built up in the early days. Up to now, comprehensive land subsidence monitoring systems in Shanghai have been set up, which consist of bedrock marks, layered marks, and GPS stations (Cui and Yuan 2015). These monitoring stations had been set up for decades of years and the monitoring systems were formed to monitor the land subsidence in Shanghai area accurately (Zhang and Gu 2000).

4.2.2 In-Site Monitoring of Land Subsidence in High-Rise Building Group

High-rise building groups are dense in Lujiazui area of Shanghai. The land subsidence is greatly affected by the construction of the dense high-rise building groups. Figure 4.2 shows four high-rise buildings, including Shanghai Merchants Tower, Jinmao Tower, BOC Tower, and SWFC. Figure 4.3 shows the location of four high-rise buildings in the map and Table 4.1 shows the basic data of four buildings.

Figure 4.4 illustrates the monitoring points around four buildings. The monitoring points were arranged along five paths, A01–A10, B01–B10, C01–A10, D01–A10, and E01–E10, respectively.

Along Huayuanshiqiao Road, at the positions of 7, 12, 20, 30, 60, 100, 150, 200, 250, and 300 m away from World Financial Center, there are 10 monitoring points: A01, A02, A03, A04, A05, A06, A07, A08, A09, and A10, respectively. Along Dongtai Road, at the positions of 7, 12, 20, 30, 60, 100, 150, 200, 250, and 300 m

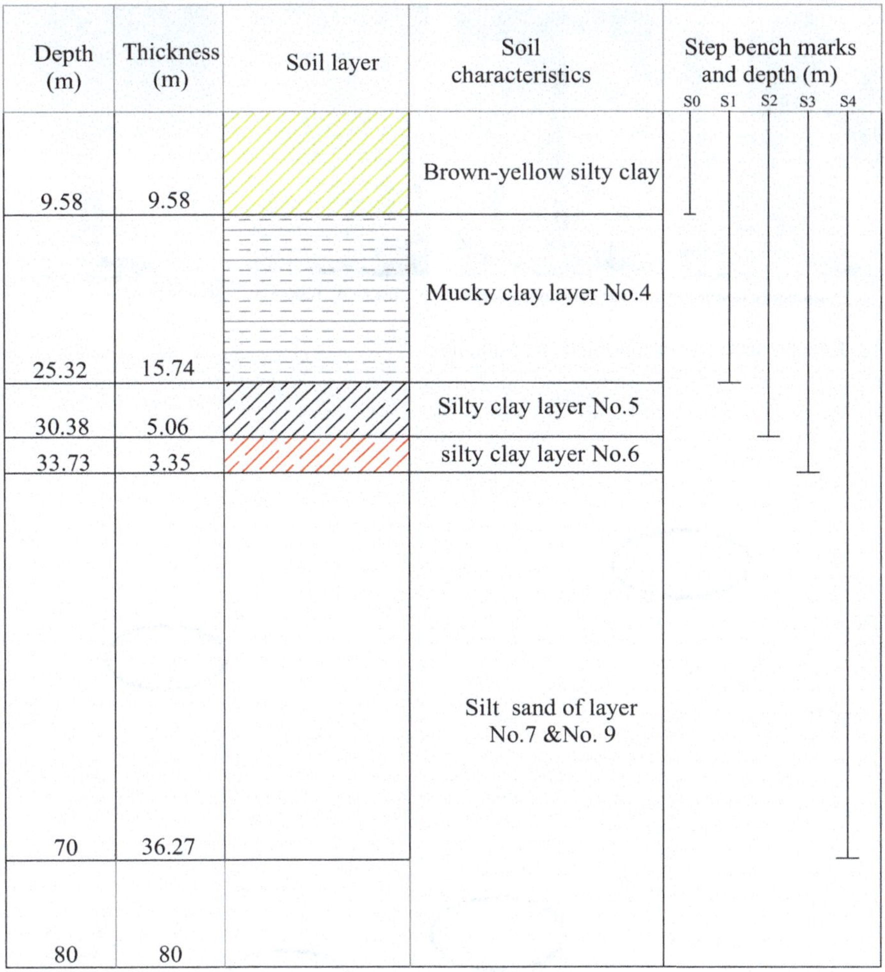

Fig. 4.1 Drill column section of soil layers in Lujiazui area

away from World Financial Center, there are 10 monitoring points: B01, B02, B03, B04, B05, B06, B07, B08, B09, and B10, respectively. The elevation of these 20 monitoring points was monitored from December 2005 to November 2008. Figures 4.5 and 4.6 illustrate the variations of land subsidence with the distance away from World Financial Center at 10 months (October 2006), 20 months (August 2007), 30 months (June 2008), and 35 months (November 2008), respectively.

Figure 4.5 illustrates the variations of land subsidence with the distance away from World Financial Center along Huayuanshiqiao Road. Point A01 experienced the maximum land subsidence, which was the nearest point to the building. With the distance increasing, the land subsidence decreased. But at Point A08 (200 m),

(a) Merchants Tower (b) Jinmao Tower (c) BOC Tower (d) World Financial Center

Fig. 4.2 Physical map of four high-rise buildings

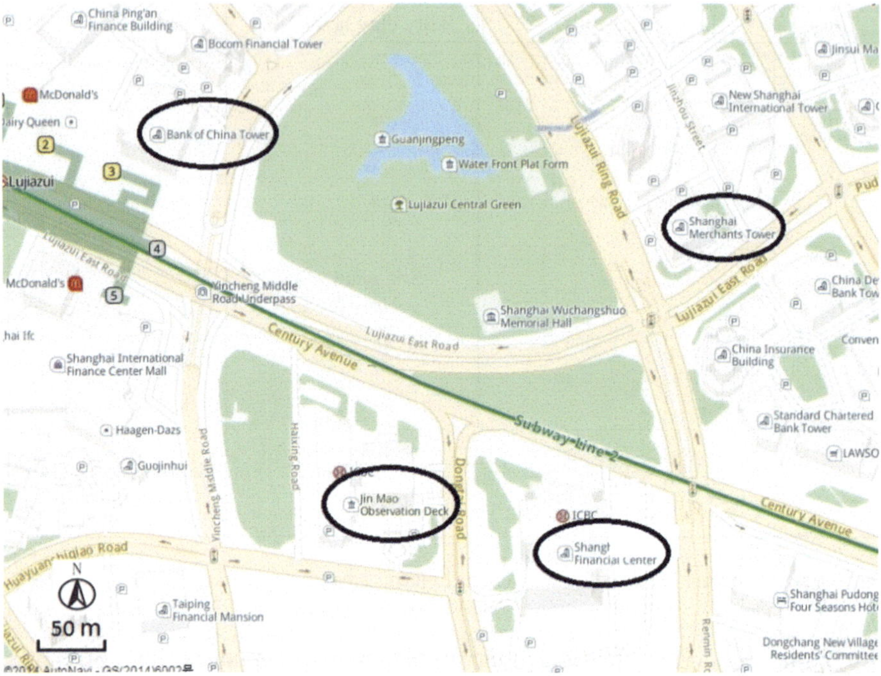

Fig. 4.3 The location of four high-rise buildings in the map

the land subsidence increased again, because Point A08 had the minimum distance from Jinmao Tower, which showed the superimposition effect of two high-rise buildings. In addition, Point A10 had greater land subsidence than Point A09, which may be affected by the traffic load of Yincheng Middle Road.

Table 4.1 The basic data of four buildings

Building name	Start time	Time of completion	Height (m)	Floor (ground/ underground)	Building area (10^3 m^2)	Pile depth (m)	Bearing layer
Merchants Tower	1992	1996	186	39/2	75	–	Layer No. 7
Jinmao Tower	1994	1999	420.5	88/3	290	83.5	Layer No. 9
BOC Tower	1997	2000	258	53/3	120	46.5	Layer No. 7
World financial center	2003	2008.8	492	101/3	381.6	78	Layer No. 9

Figure 4.5 illustrates the variations of land subsidence with the distance away from World Financial Center along Dongtai Road. The subsidence of monitoring points from B01 to B10 had the same varying regulation at 10 months, 20 months, and 30 months. However, the subsidence of monitoring points from B01 to B06 at 35 months was abnormal.

In order to further analyze the superposition effect of building loads on the land subsidence, the variations of settlement with time in the monitoring Point A08 were compared with that of the monitoring Point B08, as shown in Fig. 4.7. The variations of land subsidence of Point A08 with time was the same as that of Point B08, which mainly related to the construction process of the World Financial Center. The land subsidence at Point A08 was larger than that of Point B08 due to the effect of building load from the Jinmao Tower. At 15 months, the land subsidence of Point A08 was 113.2% larger than that of Point B08. The land subsidence caused by high-rise buildings had a remarkable superposition effect, which exceeded the allowable value for the safety of buildings. It is important to control the value of the loads in order to prevent the excessive land subsidence.

Figure 4.8 illustrates variations of subsidence with time along Path A, Path B, and Path C, respectively. At the first stage, the rate of land subsidence was small. Then the land subsidence increased and at last, the land subsidence became gentle again. Besides, the subsidence of each monitoring point from A01 to A10, from B01 to B10, and from C01 to C10 was not a simple drop with the increase of the distance from the SWFC.

From Fig. 4.8a, Point A01 had the largest subsidence and the Point A09 was the smallest. It should be noted that the subsidence at Point A08 was much larger than that at the points A05, A06, and A07, which was influenced by the stress superposition effect of both the Jinmao Tower and World Financial Center. From Fig. 4.8b, the variations of land subsidence with time along Path B were similar to those of Path A. However, a large settlement occurred at the Point B07, which might be affected by the traffic load of Lujiazui Ring Road. The land subsidence in the urban area was influenced by a variety of engineering-environment effects,

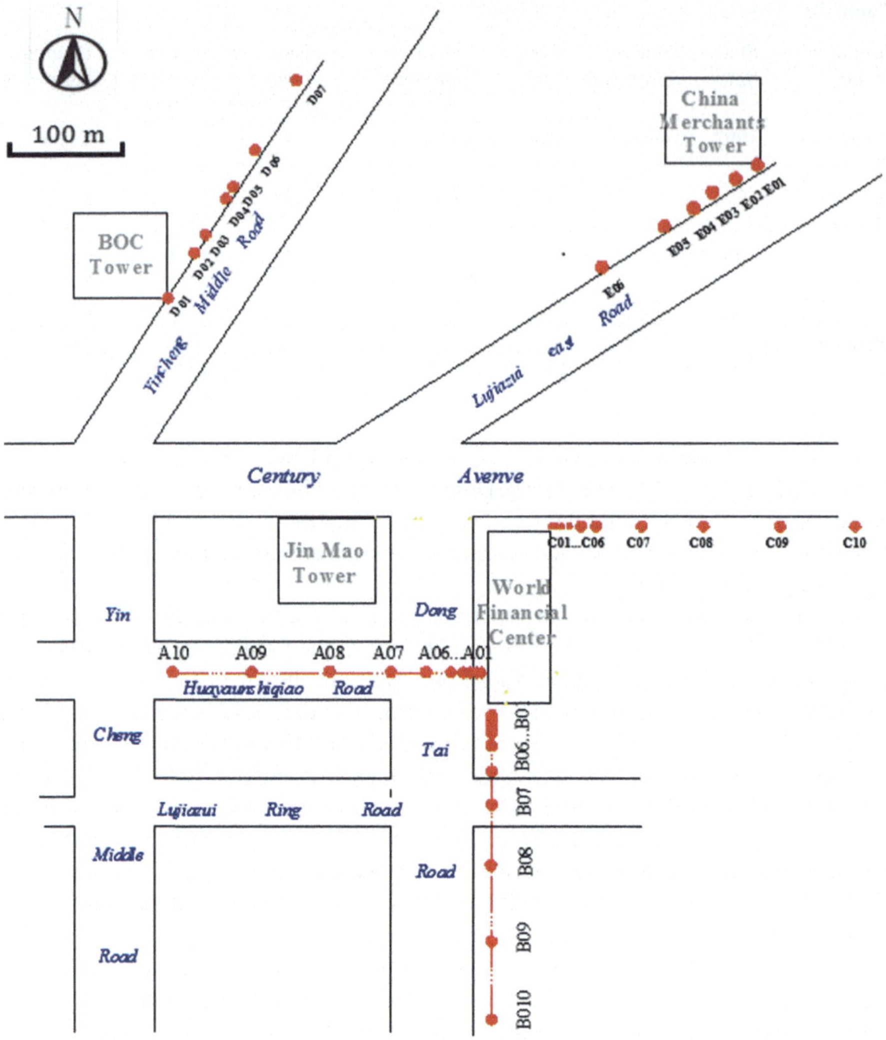

Fig. 4.4 The monitoring points around four buildings

which were characteristics of complex. From Fig. 4.8c, the land subsidence along Path C had a similar law by comparing with that along Path A and Path B.

Figure 4.9 illustrates the variations of land subsidence along Path A, B, and C with the distance away from high-rise buildings. According to the measured data, the land subsidence largely decreased with the increase in distance of the buildings. Nevertheless, the land subsidence was mutated in some monitoring points, including A08, A10, B06, B09, and C09. The monitoring points around high-rise buildings experienced larger land subsidence, whose maximum subsidence was greater than 60 mm and the maximum rate of land subsidence was up to 45 mm/a.

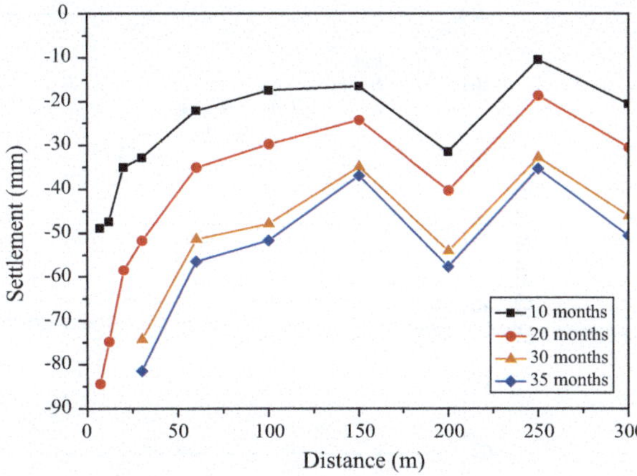

Fig. 4.5 Subsidence of A01–A10

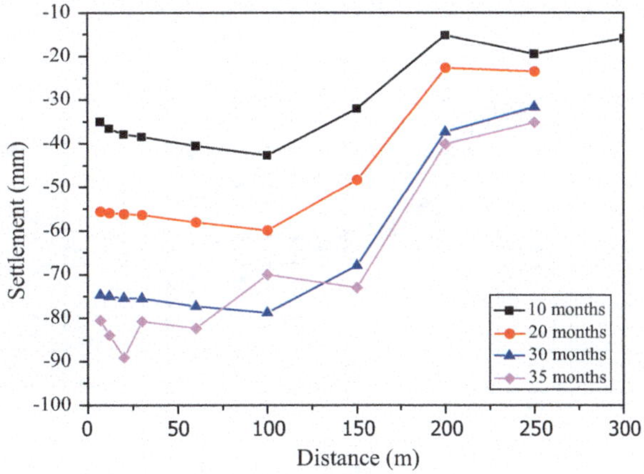

Fig. 4.6 Subsidence of B01–B10

On one hand, the rate of land subsidence in the monitor area was far greater than the average rate of land subsidence in the urban area of Shanghai during the same period. The rate of land subsidence was less than 10 mm/a at the points A10, B10, and C10, 300 m away from buildings, which was close to the average rate of land subsidence, about 6.8 mm/a, during the same period in the urban area of Shanghai. It showed that the influence range of land subsidence caused by the high-rise buildings was about 300 m, which was close to five times the width of the building foundation.

Figure 4.10 illustrates the variations of subsidence with time along Path D around BOC Tower and along Path E around Merchants Tower. The land subsidence around the BOC Tower was larger than that around the Merchants Tower. The average rate of land subsidence at Point D01 was about 25 mm/a and Point D07 experienced the largest rate of land subsidence, being 15.3 mm/a. Similarly, the largest average rate of land subsidence around the Merchants Tower was about

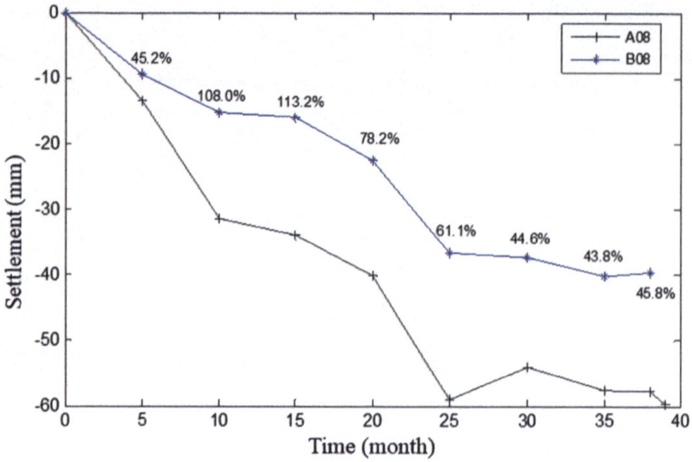

Fig. 4.7 Variations of subsidence with time (A08 and B08)

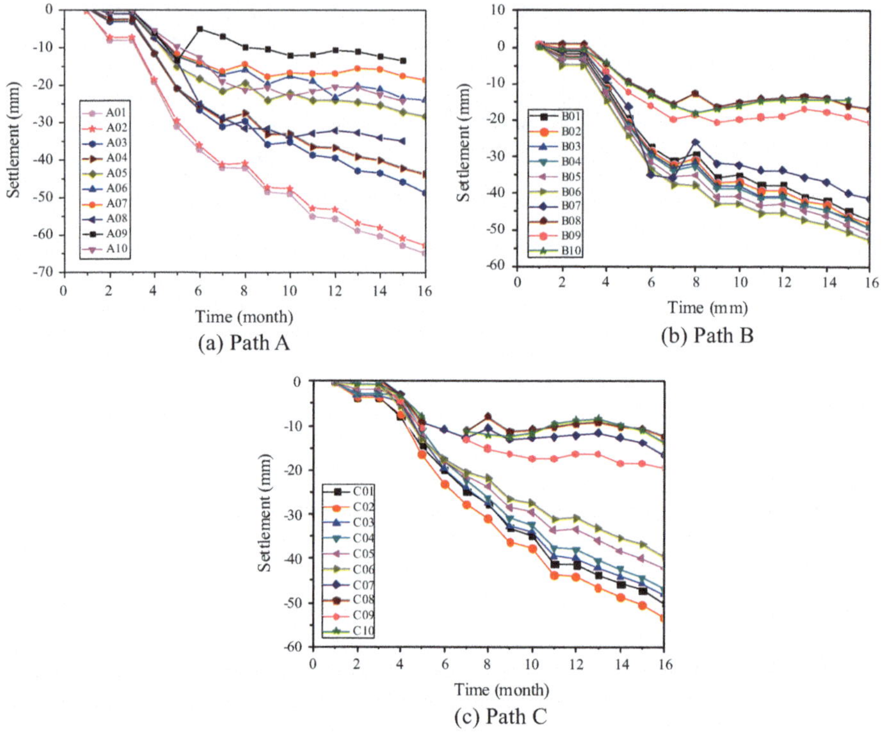

Fig. 4.8 Variations of subsidence with time along different paths

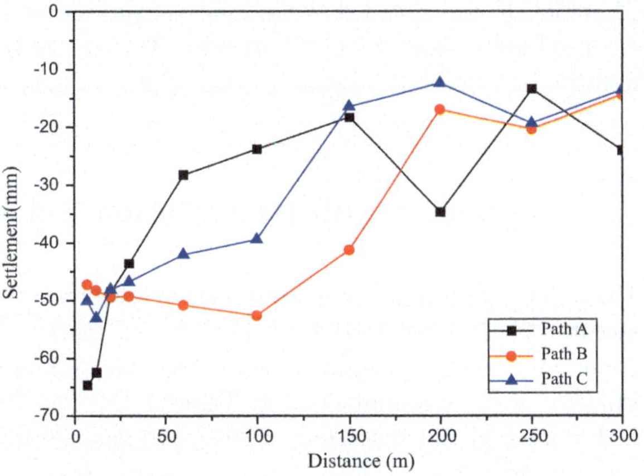

Fig. 4.9 Variations of land subsidence with distance along Paths A, B, and C

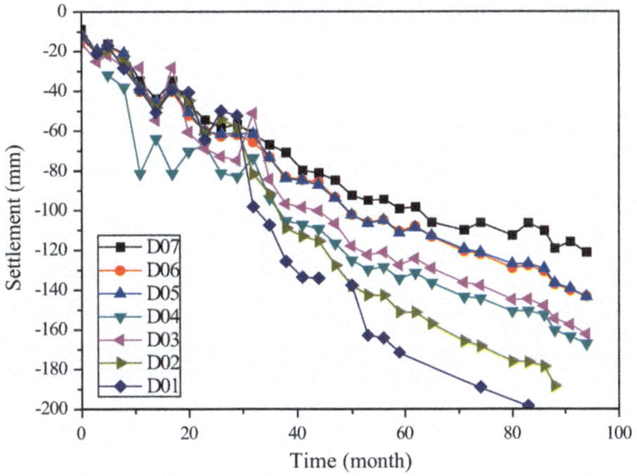

(a) Path D

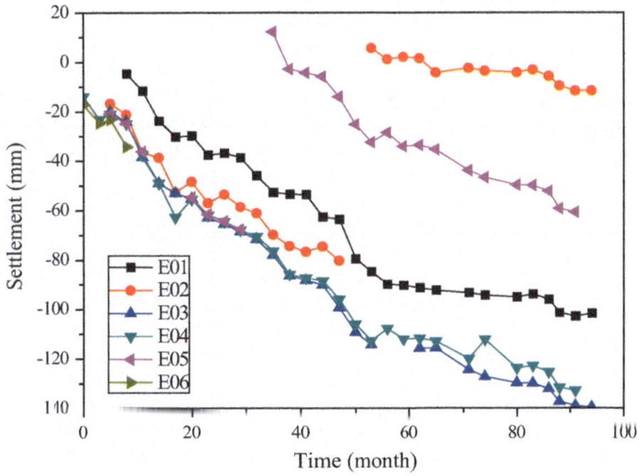

(b) Path E

Fig. 4.10 Variations of subsidence with time along Path D and Path E

25 mm/a. It was noted that the land subsidence of E02 around Merchants Tower emerged jump from 53 to 62 months. This might be due to some unknown phenomena or externally imposed stress in the vicinity of this test site.

4.3 Subsidence with Groundwater Table Changing

The deformation of layered soils and the level of groundwater were analyzed by the monitoring data obtained from the layered marks and water level holes between 2004 and 2007 in Lujiazui area. The detailed information of layered soils in Lujiazui area are summarized in Table 4.2.

Figure 4.11 illustrates the variations of land subsidence of different soil layers in Lujiazui area from August 2004 to August 2007. According to the data, the total settlement value of the upper soil minus the settlement value of lower layer was the compressive deformation of this layer.

Table 4.2 Distribution of strata in Lujiazui area

Layerwise mark	Depth (m)	Height (m)	Soil layers	Height of each soil layer (m)
S0	0.00	0–70.0	–	–
S1	9.58	9.58–70	Silty clay of layer No. 4	9.58–17.32
S2	25.32	17.32–70	Clayey soil of layer No. 5	25.32–30.38
S3	30.38	25.32–70	Clayey soil of layer No. 6	30.38–33.37
S4	33.37	30.38–70	Silt sand of layer No. 7 & 9	33.37–70.0
	70.00	–	–	–

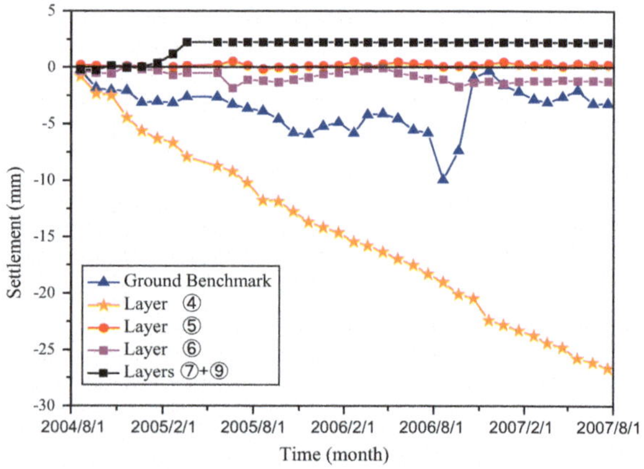

Fig. 4.11 Variations of subsidence with time for different soil layers

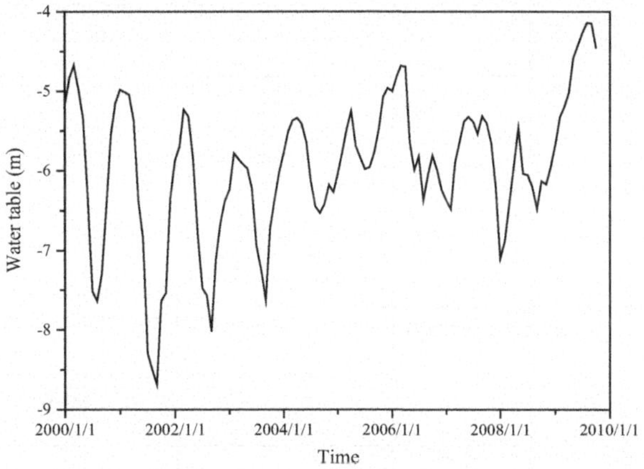

Fig. 4.12 Variations of groundwater table of the second confined aquifer with time

The mucky clay of layer No. 4 experienced the maximum settlement than other soil layers. Furthermore, the building loads increased with the continuous construction of residential area in this time period, which led to the increase in deformation of this soil. On one hand, the mucky clay of layer No. 4 buried in the depth 9.58–17.32 m bore the larger building additional load of buildings.

Figure 4.12 illustrates the variations of groundwater table of the second confined aquifer with time at the monitoring site F15. The groundwater table showed an overall rising trend with some fluctuations. As mentioned in Fig. 4.11, a slight rebound phenomenon occurred in the silt sand layer No. 7 and No. 9 during this period. The reason could be that a large number of groundwater were recharged in recent years in Shanghai. As a permeable material, particles of the sand bear less effective stress as the groundwater rises, causing a rebound in the sand layer. At the same time, settlements of the benchmark on the ground vary to some degree during this period with an average value of 5 mm, which indicates that changes of land subsidence are complex for the combining factors of the groundwater, the building loads, and the engineering-environment effect.

4.4 Numerical Simulation

4.4.1 Basic Parameters of the Model

Compared with the model tests and the in-site monitoring, the numerical simulation has the advantages of high efficiency and low cost. Land subsidence caused by high-rise buildings is a complex problem so that the analytical solution by theoretical calculation cannot be obtained. However, the numerical simulation can solve such problems. Pile raft foundations are normally used for high-rise buildings. The interaction of piles and soils has been theoretically analyzed and modeled by many

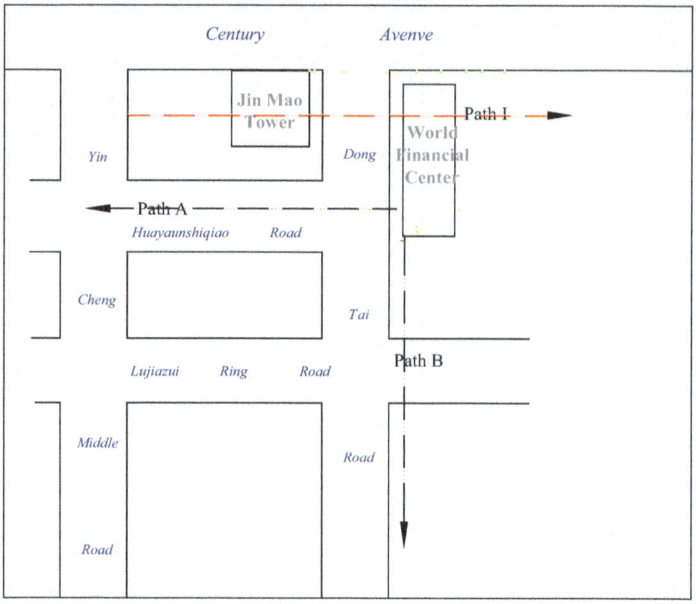

Fig. 4.13 Layout of buildings in numerical simulation model

scholars. However, it is very difficult to accurately calculate the land subsidence caused by high-rise buildings, because there exists the elastic, plastic, and viscous plastic deformation in the soft soil.

Due to the complexity of the actual project, it is difficult to accurately simulate the land subsidence in the study area. So, only two super high-rise buildings were selected, which were Jinmao Tower and SWFC and the numerical simulations were conducted to study the land subsidence caused by two closely spaced high-rise buildings. Figure 4.13 illustrates the layout of buildings in the numerical simulation model. In addition, three representative paths Path A, Path B, and Path I were selected to obtain variations of land subsidence with different positions of buildings themselves and their surrounding soils provide. The physical and mechanical properties of soils in the area of Jinmao tower and SWFC are summarized in Tables 4.3 and 4.4. Due to the complex geological conditions of the actual conditions, the simplification of the soil layers in this numerical simulation was conducted. The basic data of piled raft foundations for two buildings are summarized in Table 4.5.

4.4.2 Analysis of Numerical Simulation

The material of the raft was considered as the linear elastic material and the soils were treated as the elastoplastic material. Expressions of the shear force and the

Table 4.3 Physico-mechanical properties of soils for Jinmao Tower

Layer numbers	Soil name	Height (m)	Elevation on roof (m)	Water content w (%)	Gravity γ (kN m^3)	Void ratio e	Permeability k ($\times 10^{-4}$ cm s^{-1})	Cohesion c (kPa)	Internal friction angle $\phi/(^\circ)$
1	Filled soil	0.9	+3.800	–	–	–	–	11.0	17.0
2	Silty clay	2.715	+2.900	35.3	18.5	1.00	0.511	17.0	20.8
3	Muddy silty clay	4.2	+0.725	39.6	18.1	1.11	1.77	11.5	22.0
4	Mucky clay	9.72	−3.475	49.0	17.3	1.37	0.0164	14.0	13.5
5	Silty clay	8.59	−13.195	34.4	18.5	0.98	0.00133	13.0	20.0
6	Silty clay	3.215	−21.785	23.0	20.1	0.67	–	51.0	21.0
7-1	Sandy clay	6.94	−25.00	31.2	18.6	0.91	–	4.3	32.7
7-2	Silty fine sand	28.32	−31.94	26.9	18.9	0.80	–	0.0	33.6
8	Sandy silt	–	–	32.1	18.5	0.93	–	–	–
9-1	Sandy silt	–	–	28.9	18.9	0.84	–	–	–

Table 4.4 Physico-mechanical properties of soils for SWFC

Layer numbers	Soil name	Height (m)	Water content w (%)	Gravity γ (kN m^3)	Void ratio e	Permeability k ($\times 10^{-4}$cm s^{-1})	Cohesion c (kPa)	Internal friction angle $\phi/(^\circ)$
1	Filled soil	2.9	–	–	–	–	–	–
2	Clay silty clay	1.4	36.1	18.5	1.02	0.00737	16.0	12.0
3	Muddy silty clay	4.2	39.7	18.1	1.10	1.10	6.0	20.3
4	Mucky clay	10.4	48.6	17.3	1.36	0.00824	10.0	10.0
5	Silty clay	7.6	32.7	18.8	0.92	0.0234	10.0	14.0
6	Silty clay	3.9	23.2	20.0	0.68	0.707	36.0	12.7
7-1	Sandy clay	12.6	30.0	19.1	0.83	5.36	2.0	25.8
7-2	Silty fine sand	20.8	28.2	19.7	0.72	8.21	0.0	26.8
7-3	Sandy silt	9.3	30.1	19.3	0.82	–	2.0	25.0
9-1	Silty clay silt	5.7	27.7	19.5	0.76	–	2.0	25.0
9-2	Medium-coarse sand	14.5	19.1	20.9	0.53	–	2.0	25.0
9-3	Silty fine sand	54	24.3	20.0	0.68	–	–	–
10	Silty clay	18.7	26.3	19.8	0.75	–	–	–

Table 4.5 Basic data of piled raft foundations for two buildings

Building name	Height (m)	Number of floors	Raft thickness (m)	Buried depth (m)	Number of piles	Pile length (m)	Total load ($\times 10^6$ kN)	Foundation area (m^2)	Pressure raft foundations (kPa)
Jinmao Tower	420.5	88	4.0	19.65	429	83	3.0	3519	852.5
SWFC	492	101	4.5	18.54	1177	79	4.4	6200	709.7

shear displacement of the soils and piles are obtained as the form of penalty functions according to Mohr–Coulomb yield criterion. The simplification of the soil layers in this numerical simulation was conducted. The specific parameters of the model are summarized in Table 4.6.

The boundary conditions are as follows:

(1) Physical boundaries: displacements of lateral and bottom faces of the model do not exist.
(2) Mechanical boundaries: the top surface is free for water in soils to go while the sides and the bottom face are undrained.
(3) Initial conditions: freedom of pore water pressure is zero when the pore water is immobile. Subsidence of soil mass is steadily stated, so the pore water pressure of all nodes in the model is also equal to zero.

Figure 4.14a illustrates the final subsidence along Path A on Huayuanshiqiao Road. The maximum settlement of the numerical simulation was 450 mm, which was much larger than that of the in-site monitoring. But the in-site monitoring was carried out only for 30 months and its maximum settlement was 84 mm. So, the land subsidence in Shanghai area has characteristics of long duration and strong hysteresis. The soil around the high-rise building foundations suffered the drainage consolidation and secondary consolidation phenomenon for a period of time after the building was completed. According to the monitoring data (Ma et al. 2006), the instantaneous settlement of soft soil accounted only for 10–30% of total settlement under the action of external loads in Shanghai. The time of the foundation settlement of buildings might be more than 10 years.

It should be noted that the land subsidence decreased significantly with the increase in distance of the SWFC. Comparing the in-site monitoring data as shown in Fig. 4.5, two settlement curves had a turning near the Point A08 in some degree, which was due to the influence of the load of Jinmao Tower. The additional stress at

Table 4.6 A numerical model of material parameters

Materials	Elastic modulus (kPa)	Poisson's ratio μ	Permeability k ($\times 10^{-4}$ cm s^{-1})	Gravity (kN/m^3)	Cohesion c (kPa)	Internal friction angle (°)
Pile	200,000	0.2	–	–	–	–
Brown-yellow silty clay	150	0.35	1.435	18.1	9	21
Mcuky clay layer No. 4	50	0.4	0.01232	17.3	12	12
Silty clay layer No. 5	110	0.35	0.01237	18.7	14	17
Silty clay layer No. 6	110	0.35	0.707	20.0	44	17
Silt sand layer No. 7 & No.9	500	0.3	8.21	19.3	1	28

Fig. 4.14 Land subsidence at different paths

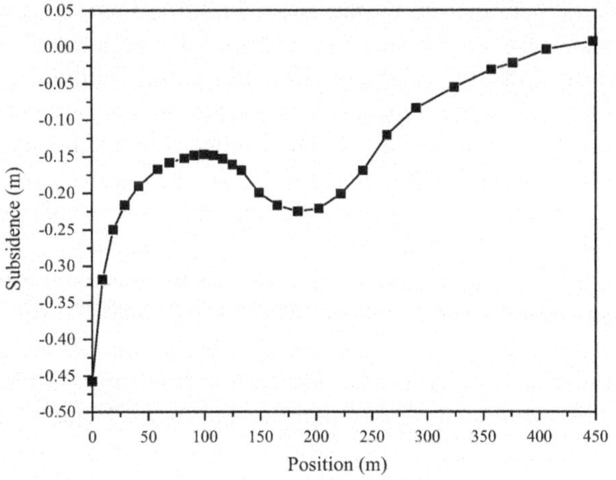

(a) Path A

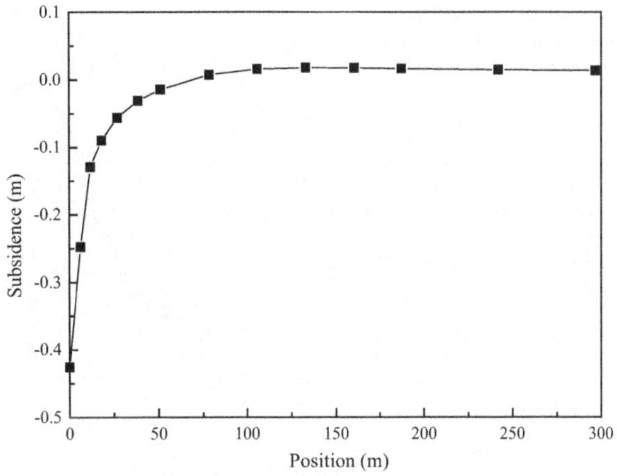

(b) Path B

the Point A08 was affected by the superposition of two building loads. It manifested that land subsidence caused by high-rise buildings was undermined by the multiple factors and the complex variation with the different positions.

Figure 4.14b illustrates the final subsidence along Path B on the Dongtai Road. Due to the ideal condition of the numerical simulation, the land subsidence along Path B in the numerical simulation was different from that of the in-site monitoring, which resulted from some unknown phenomena or externally imposed stress in the vicinity of the test site, such as B07 been experienced by the vehicle dynamic load. It indicated that with the distance of the building foundation increasing, the subsidence decreased rapidly and the land subsidence was basically zero at a distance of 100 m. Although the foundation and its surrounding soil experienced a larger settlement caused by the larger building load, the impact ranges of the settlement were also very limited. In addition, the settlement was affected by the foundation form and the weight of the surrounding buildings. For the rectangular or strip foundation, the impact of the settlement was generally not more than two times the width of the foundation. The excessive land subsidence can not only threaten the safety of the building itself but also create serious hazard to its surrounding project.

Figure 4.15 illustrates the variations of the additional stress with the different positions. When the load of Jinmao Tower was applied, the maximum additional stress was generated at the center of the base and gradually decayed to zero along the radius of the 150 m. After the Jinmao Tower and SWFC loads were both applied, the additional stresses in the bottom of the two building foundation center had two extreme points. Meanwhile, the superposition effect of stresses existed in the central area of two high-rise buildings. Figure 4.16 illustrates the variations of

Fig. 4.15 Variations of additional stress with positions (Path I, 100 m soil depth)

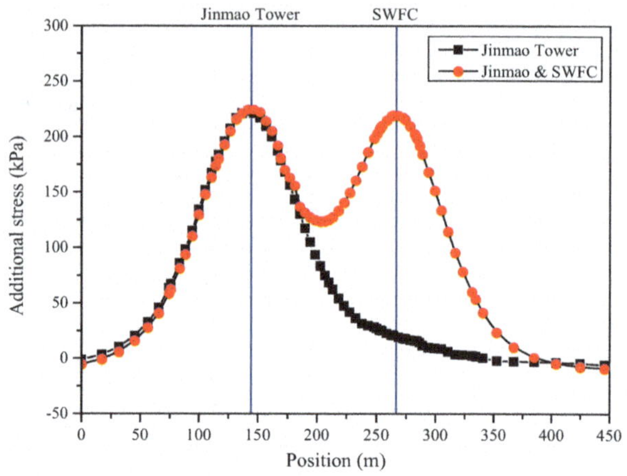

Fig. 4.16 The subsidence of Path I

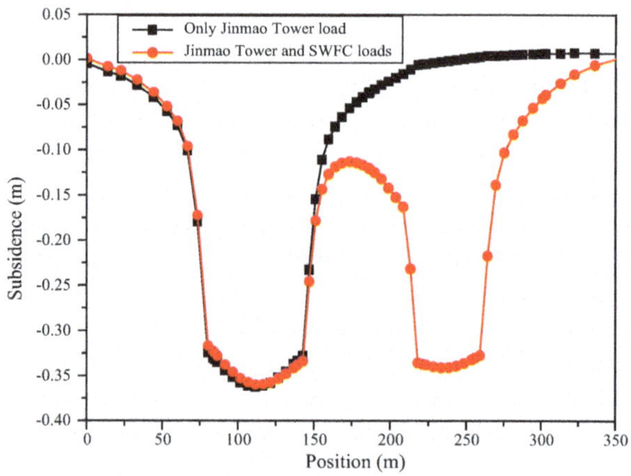

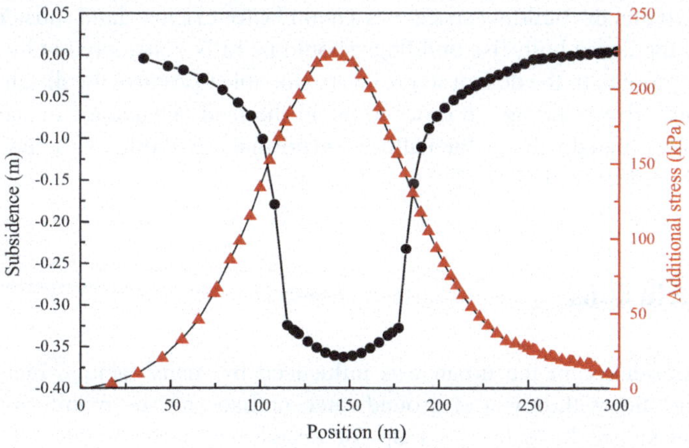

(a) Jinmao Tower load

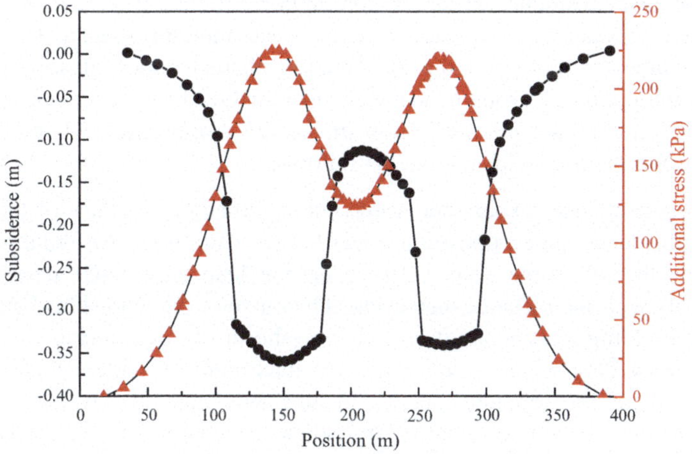

(b) Jinmao Tower and SWFC loads

Fig. 4.17 Variations of subsidence and of additional stress at different loading conditions

land subsidence with different positions along Path I. The building itself experienced the maximum subsidence. The differential subsidence occurred in the building foundation and its surrounding ground. Comparative analysis of Figs. 4.15 and 4.16 showed that the distribution of soil settlement was determined by the distribution of the additional stress of the foundation under the flexible pile cap.

Figure 4.17 illustrates the variations of the subsidence and the additional stress with the different positions under the load of Jinmao Tower and the loads of Jinmao Tower and SWFC, respectively. The magnitude of the additional stress of the soil was the decisive factor to determine the final settlements of the soil. On the other hand, the superimposition effect of the high-rise buildings in the central area was obvious.

This indicated that the building space was a main factor to cause land subsidence. The influence of the dense high-rise buildings should be fully considered in the practical phase of the project. In the design stage, it is reasonable to control the distance among the dense high-rise buildings in order to avoid the land subsidence of the regional ground over estimated value, which affects the normal use of other structures, such as the roads, the underground pipelines, and so on.

4.5 Conclusions

The land subsidence in the urban was influenced by many factors, including the building load, the withdrawal of groundwater, and so on. The monitoring stations for land subsidence have been set up for decades of years in Shanghai, which consist of the bedrock marks, the layered marks, the water level holes, etc. In this chapter, the land subsidence caused by four high-rise buildings which are located in Lujiazui area of Shanghai was studied by the in-site monitoring data. The changing of the groundwater table and the deformation characteristics of soils under the different exploitation of groundwater were also analyzed by the in-site monitoring data. Based on the data obtained from the in-site monitoring and the numerical simulation, the main conclusions are as follows.

(1) Building itself may experience a maximum settlement because of the remarkable subsidence superimposition effect between the high-rise buildings which exceeds the allowable values. However, the land subsidence decreases dramatically with the distance increasing. The range of the land subsidence caused by the building load is about 300 m from the in-site monitoring.
(2) The mucky clay of layer No. 4 has the maximum settlement than other soil layers. A slight rebound occurs in the silt sand of layers No. 7 and No. 9 because a large amount of groundwater recharge has been conducted in Shanghai in recent years. As a permeable material, particles of the sand bear less effective stress as the groundwater rises, causing a rebound in the sand layer.
(3) The average settlement is 5 mm except some fluctuations, which indicates that the influence factors of land subsidence are very complex, including the withdrawal of groundwater, the building loads, and the engineering-environment effect, etc.
(4) The surrounding ground of high-rise buildings suffers obviously superimposed subsidence by the additional stress, which causes a larger ground settlement than the estimated value. Therefore, in the design stage, it is reasonable to control the distance between the dense high-rise buildings in order to avoid the settlement of regional ground over the allowable value and affect the normal use of other structures.

References

Cui ZD, Tang YQ (2011) Microstructure of different soil layers caused by the high-rise building group in Shanghai. Environ Earth Sci 63:109–119

Cui ZD, Yuan Q (2015) Study on the settlement caused by the Maglev train. Nat Hazards 75:1767–1778

Gong SL (1998) Effects of urban construction on the land subsidence in Shanghai. Chin J Geol Hazard Control 9(2):108–111

Ma GH, Chen JJ, Wang JH, Zhou HB (2006) Long-term settlement of multi-story buildings on soft foundation in Shanghai. Rock Soil Mech 27(6):0991–0994. (in Chinese)

Jie YX, Gao Y, Li GX (2007) Analysis on the land subsidence induced by city construction. Geotech Eng Tech 21(2):78–82

Tang YQ, Cui ZD, Wang XH (2007) Preliminary research on the land subsidence induced by engineering environmental effect of dense high-rise building group. Northwest Seismol J 29 (2):105–108

Tang YQ, Cui ZD, Wang JX, Yan XX (2008) Model test study of land subsidence caused by high-rise building group in Shanghai. Bull Eng Geol Environ 67:173–179

Zhang AG, Gu WD (2000) Design principle and construction technique of ground subsidence monitoring bench mark in Shanghai. Explor Eng 5:67–70 (in Chinese)

Zhang Y, Xue YQ, Li QF (2003) Current prominent subsidence layer and its deformation properties in shanghai. Hydrogeol Eng Geol 30(5):6–11

Chapter 5
Centrifuge Modeling of Land Subsidence Caused by the High-Rise Building Group

5.1 Introduction

In the coastal urban area of China, the groundwater extraction is the primary cause of land subsidence. But from the 1960s, the withdrawal of groundwater was controlled reasonably and especially, from the end of the 1970s, the pumping of groundwater was strictly controlled in the urban area in Shanghai. The quantity of water recharged into the subsurface was always greater than that of the pumping and the extracted aquifers were gradually adjusted. As a result of these measures, the subsidence caused by pumping and recharging was kept smooth and gentle in the urban area (Cui et al. 2010a). During the 1990s, however, with the development of the economy, a variety of municipal works and high-rise buildings were constructed and the subsidence appeared to accelerate in Shanghai (Cui et al. 2010b).

With an increasing number of high-rise buildings in major cities worldwide, land subsidence induced by the interaction between closely spaced buildings becomes a major challenge to many researchers. The mechanism of land subsidence caused by interaction of high-rise buildings was first studied by the model test (Tang and Cui 2008). The land subsidence caused by high-rise buildings was affected by the construction sequence of the buildings and the land subsidence of the points near to the former building was larger than that of those near to the latter building; the central area of the group of buildings experienced the maximum subsidence and the points within one time the width of the foundation from the center of the building had the second largest subsidence. But the general model test cannot satisfy the requirement that the model had the same stress level as the prototype and the centrifuge model test was conducted by Cui et al. (2010a). The central area of the building group had larger subsidence and the subsidence superimposition effect was obvious. The buildings distance was smaller and the subsidence superimposition effect was more obvious.

This chapter discusses the engineering-environmental effect of high-rise building group on land subsidence under the typical geological subsurface of Shanghai by

Z.-D. Cui, *Land Subsidence Induced by the Engineering-Environmental Effect*,
https://doi.org/10.1007/978-981-10-8040-1_5

the centrifuge model tests. The problems studied include the deformation characteristics of different soil layers, the influence of high-rise building group on land subsidence of its central and circumjacent areas, the land subsidence affected by the different building distances, and the stress variation of soil layers due to the engineering-environmental effects of the high-rise building group.

5.2 Determination of Bearing Capacity of Single Pile

It is of importance to determine the bearing capacity of single pile in the centrifuge model test which is the precondition to correctly study the mechanism of land subsidence. The T-bar penetration test was conducted in flight to measure the shear strength of the soft soil in this chapter.

5.2.1 In-Flight T-Bar Penetration Test

Centrifuge modeling is a useful tool for the investigation of geotechnical problems (Powrie and Daly 2007; Geddes 1966) because of its ability to reproduce the same stress levels in a small-scale model, as those present in the full-scale prototype.

5.2.1.1 The Geotechnical Centrifuge

The geotechnical centrifuge used for the model tests has a single arm with the nominal radius of 3 m. The capacity of the centrifuge is 150 g·t and its maximum acceleration can reach 200 g. The strongbox used for the model tests has dimensions of 500 mm × 800 mm × 550 mm in width, length, and height. Three side walls and the bottom of the container are made of high strength stainless steel plates to protect against possible corrosion and to reduce the friction between the soil and the container surface. The front wall of the container is made of a 40-mm-thick Plexiglas plate.

5.2.1.2 Soil Property

According to the geological and hydrogeological background in Shanghai, the soil layers from the top downward are brown-yellow clay layer, silty clay layer, silt sand layer, clayey soil layer, and silt sand layer. The foundation soil of the model test is silty clay of layer No. 4, silt sand of layer No. 7, and clayey soil of layer No. 8, respectively, obtained from the site in Shanghai. The soil properties are summarized in Table 5.1.

Table 5.1 Physical and mechanical properties of soils for the centrifugal model test

Soil	Water content (%)	Bulk density (kN/m^3)	Initial void ratio	Compression modulus (MPa)	Cohesion (kPa)	Internal friction angle (°)
Silty clay of layer No. 4	43.58	18.89	1.075	2.48	12.0	12.5
Silt sand of layer No. 7	9.95	13.81	1.142	9.04	5.0	24.8
Clayey soil of layer No. 8	38.79	18.88	1.001	2.13	15.8	14.5
Silt sand of layer No. 9	9.95	13.81	1.142	9.04	5.0	24.8

5.2.1.3 Preparation of Soil Layers

To minimize side friction, the container wall was covered with a thin layer of smooth plastic membrane. The procedures of preparing soil layer were as follows:

(1) The silt sand obtained from the site was used to construct the bottom sand layer (layer No. 9), 50 mm in thickness.
(2) Layer No. 8, 150 mm in thickness, was constructed by the clayey soil obtained from the site. This layer was divided by several sublayers. After the first sublayer was constructed, the second sublayer was subsequently constructed.
(3) Third, layer No. 7, 75 mm in thickness, was also constructed by silt sand.
(4) The silty clay was used to construct layer No. 4, 200 mm in thickness. Its construction way was as same as that of layer No. 8.
(5) Lastly, 25-mm-thick upper layer was constructed.

5.2.1.4 Undrained Shear Strength Measured by in-Flight T-Bar Penetration Test

A T-bar penetrometer (Fig. 5.1) was used to measure the undrained shear strength of the silty clay in flight. Figure 5.2 shows the calibration curve of the T-bar penetrometer (Cui 2012).

After installing T-bar penetrometer at 1 g, the model was accelerated to 200 g for about 3 h till the pore water pressure became stable. Figure 5.3 illustrates the time subsidence curves of soil layers during the centrifuge consolidation. Then, the T-bar penetration test was conducted in flight. Figure 5.4 illustrates the measured undrained shear strength of soft-clay layer with depth. Immediately after swinging down, vane shear test was conducted at 1 g to measure the undrained shear strength of the silty clay. The shear strength measured by vane shear immediately after centrifuge test is summarized in Table 5.2.

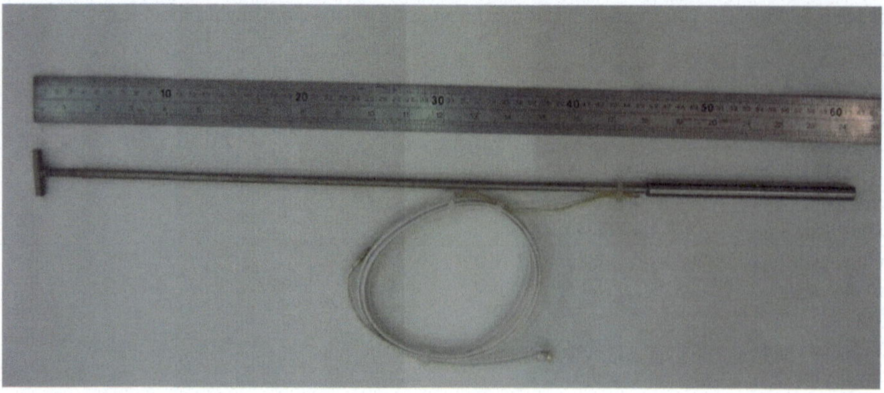

Fig. 5.1 The T-bar penetrometer used in the test

Fig. 5.2 The calibration curve of T-bar

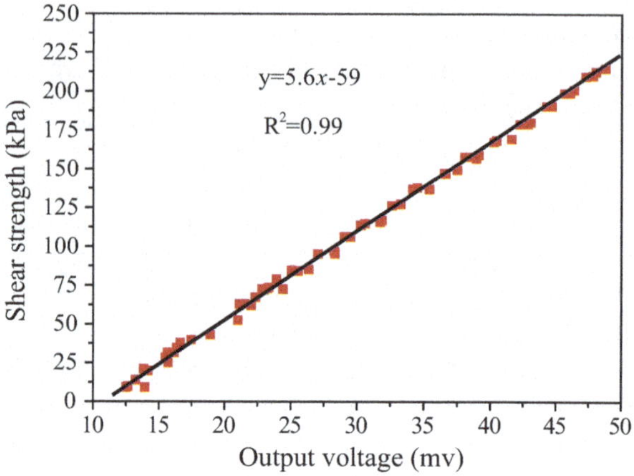

Fig. 5.3 Variations of subsidence of soil layers with time under 200 g consolidations

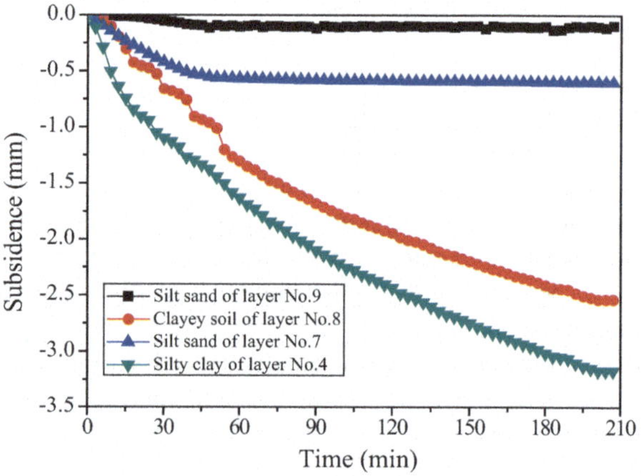

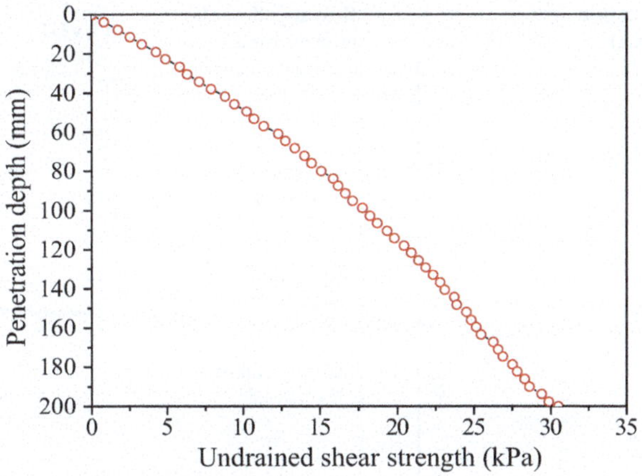

Fig. 5.4 Measured undrained shear strength with depth by in-flight T-bar test

Table 5.2 Shear strength measured by vane shear immediately after centrifuge test

Penetration depth (mm)	50	100	150	200
Shear strength (kPa)	5	8.5	11	14

Comparing the shear strength in Table 5.2 with that in Fig. 5.4, the undrained shear strengths by the vane shear test immediately after swinging down are 49.0% of that by the in-flight T-bar test at 50 mm depth, 47.2% at 100 mm depth, 45.8% at 150 mm depth, and 46.7% at 200 mm depth, respectively.

5.2.2 *Centrifuge Model Test for the Bearing Capacity of Single Pile*

5.2.2.1 Single Pile

Pile foundations are the normal type of foundation for high-rise buildings in Shanghai. The prototype plan dimensions of each building were 20 m × 20 m, with 22 floors reaching a height of 65 m. The scale of the test is 1:200. In order to simplify the model, steel bar was used to make the 230 mm model single pile and steel plate was used to make the 100 mm × 100 mm model pile cap). Figure 5.5 illustrates the model section of single pile centrifuge model test and the bearing stratum of the pile tip was the silt sand of layer No. 7. The linear variable differential transformer was installed on the pile cap in order to measure the subsidence of the building.

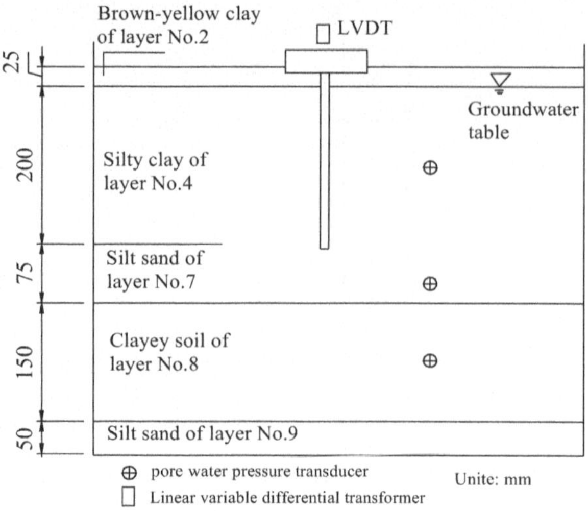

Fig. 5.5 Model section with instrumentation used

5.2.2.2 Bearing Capacity of the Single Pile

The loading pattern of the centrifuge model test for bearing capacity of the single pile is different from that of the routine test with step loading. In the centrifuge model test for bearing capacity of the single pile, the step loading pattern was conducted by increasing the acceleration of the centrifuge. The acceleration of the centrifuge a was calculated by the rotational speed ω to the centrifuge used in this test,

$$a = 3\left(\frac{\omega}{29.894}\right)^2 \tag{5.1}$$

Six steps of stable accelerations were used in this test, including 5, 10, 15, 25, and 30 g and each step was running 20 miniatures. The variations of the subsidence with the acceleration are shown in Fig. 5.6a–c. Figure 5.6d illustrates the variations of the subsidence with the running time of the centrifuge at the certain acceleration. From Fig. 5.6d, the curves between subsidence and time at each g level were analyzed and the acceleration for the bearing capacity of single pile was determined to be 25 g. The bearing capacity of the single pile was 49 kN/m^2.

In order to determine the number of the piles of the high-rise building, the additional stress (p) is given,

$$p = \gamma_G \alpha \omega N_s, \tag{5.2}$$

where ω is the vertical load of unit area and its value is 12–14 kN/m^2 for the portal frame construction; N_s is the floor number of the building; α is the magnification factor of the axial force considering the earthquake and its value is 1.05–1.10 for

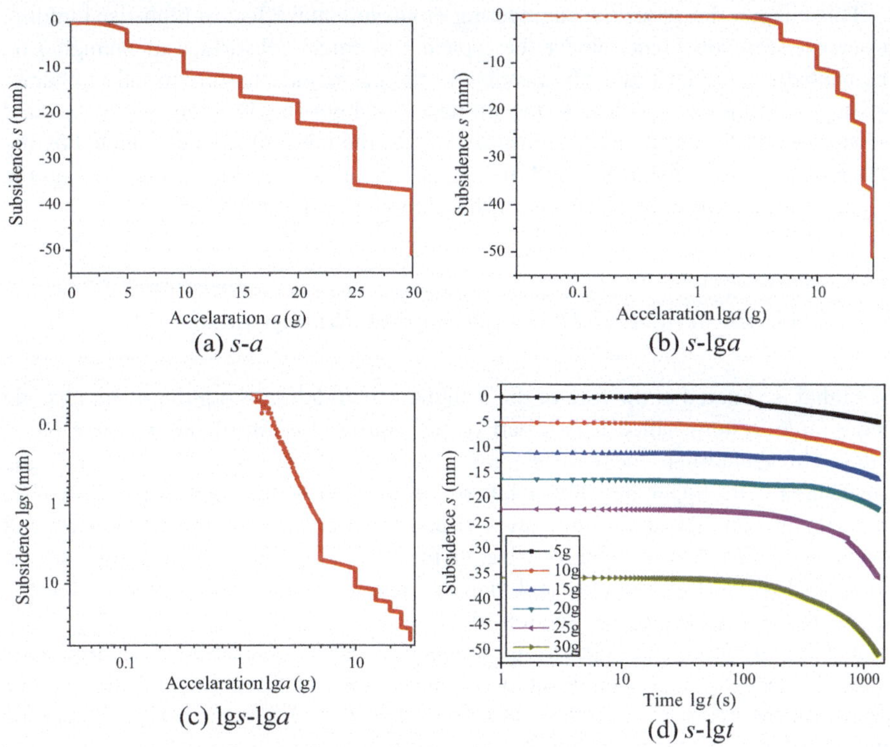

Fig. 5.6 Variations of subsidence with acceleration and time

seven seismic intensity; γ_G is the partial coefficient of the vertical load and its value is 1.2.

To the prototype building with 22 floors, $p = 1.2 \times 1.05 \times 14 \times 22 = 387$ kN/m^2. Considering the safety coefficient with its value being 2, the number of the piles of the high-rise buildings can be obtained, being 16.

5.3 Centrifuge Modeling Land Subsidence Caused by the High-Rise Buildings

It is difficult to monitor the subsidence caused by a large number of high-rise buildings for a long time and the environment of the site changes greatly with the lapse of time. The measured data are affected by many factors, the measurement of which is not available. The centrifuge model tests are applicable to study the land subsidence caused by the engineering-environmental effect of high-rise building group.

This chapter discusses the engineering-environmental effect of high-rise building group on land subsidence under the typical geological subsurface of Shanghai by the centrifuge model tests. The problems studied include the deformation characteristics of different soil layers, the influence of high-rise building group on land subsidence of its central and circumjacent areas, the land subsidence affected by the different building distances, and the stress variation of soil layers due to the engineering-environmental effects of the high-rise building group.

5.3.1 Geological and Hydrogeological Background

Shanghai is situated at the fore of the Yangtze River delta. Its aquifer is not a single and simple system, since it is a part of the aquifer system of the Yangtze River delta. The Quaternary deposit is thick and generally reaches 250–300 m in the urban area. The upper and lower layers are divided at the depth of 150 m. The upper layer consists mainly of gray soil and is composed of soft-plastic clay and sandy soil. The lower layer consists mainly of soil with mixed colors and is composed of hard-plastic clay and alternate layers of sand and gravel. Figure 5.7 shows the typical geological section of Shanghai.

Municipal works and high-rise buildings have been constructed on a large scale since the 1990s. Land subsidence accelerated again in Shanghai and the average yearly subsidence was more than four times that of the previous period (the 1970s and the 1980s). As the pumping of groundwater was strictly controlled in the urban area of Shanghai and the quantity of back-pouring of water was always greater than that of the pumping, the subsidence caused by pumping and back-pouring was kept smooth and gentle in the urban area. Under this circumstance, the effect of the large-scale construction in the urban area was taken into consideration seriously.

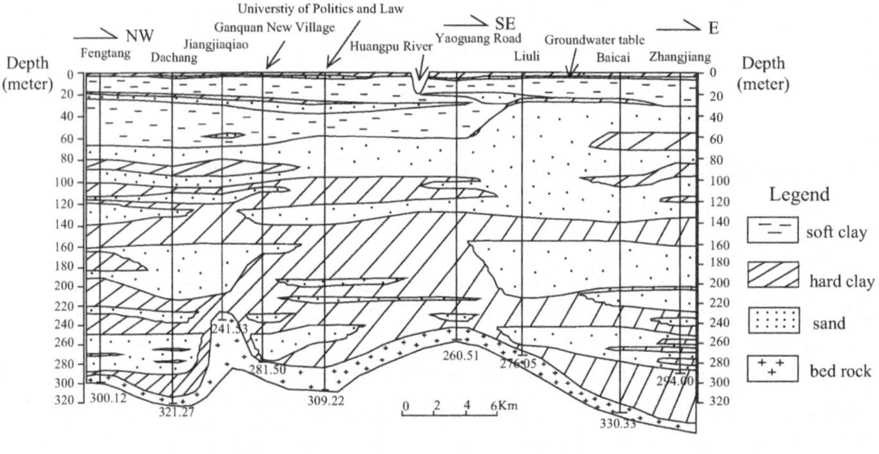

Fig. 5.7 Hydrostratigraphy along the cross section of Fengtang–Yaoguang Road, Zhangjiang

In recent years, according to the quantities of water pumped and recharged, the deformation data of soil strata indicated subsidence mainly below the depth of 70 m. The soil deformation above this depth was mainly caused by engineering works. Within the soil strata above the depth of 70 m, there are three thick layers of soft soil, which is the typical area of soft soil foundation, as summarized in Table 5.3.

5.3.2 *Centrifugal Model Tests*

5.3.2.1 Soil Property and Pile

The soil properties are summarized in Table 5.1. The model used pile foundations, as this is the normal type of foundation for high-rise buildings in Shanghai. However, it is not intended to study the characteristics of pile foundations and they are used in the model simple to take the building load. In order to simplify the model, steel bars are used to make the 230 mm model piles and steel plates are used to make the 100 mm × 100 mm model pile caps.

5.3.2.2 Preparation of Soil Layers

To minimize side friction, the container wall was covered with a thin layer of smooth plastic membrane. The procedures of preparing soil layer were as same as those in Sect. 5.2.1.3.

After soil layers were constructed, the model was consolidated at 200 g for about 3 h. Figure 5.8 illustrates the time subsidence curves of soil layers during the centrifuge consolidation.

5.3.2.3 Instrumentation and Test Procedures

The earth pressure cells (GTI-E201-500KPS) and pore pressure transducers (KYJ-31) were installed to study the characteristics of the soil pressure and excess pore water pressure of the different soil layers, shown in Fig. 5.9. Linear variable differential transformers (RMWY-50) were installed on the pile cap of one building and at various points in the foundation soils in order to measure the building and land subsidence as well as the deformation of different soil layers.

The model consisted of four high-rise buildings, using a scale of 1–200. The original plan dimensions of each building were 20 m × 20 mm, with the 22 floors reaching a height of 65 m. The bearing stratum of the pile tip was the silt sand of layer No. 7.

Table 5.3 Geological layers (above 70 m)

Engineering geology layer	The position of aquifer	The depth of the roof (m)	Thickness (m)	Layer number	Second layer	Soil characteristics
Alluvium of layer No. 1	–	–	0.00–7.00	I_1	I_1	Brown-yellow silty clay
Sand layer of layer No. 2	Underwater	0.90–7.00	1.50–27.20	I_2	I_2	Gray silt with sandy silt
Hard soil of layer No. 3	–	1.00–5.50	1.00–11.50	I_3	I_3	Brown-yellow, dark green clayey soil
Soft soil of layer No. 4	–	About 3.00	12.00–20.00	I_4	I_4^1	Gray mucky silty clay with thin silt layer in-between
					I_4^2	Gray mucky clay
Soft soil of layer No. 5	–	18.00–22.00	Several m–30.00	I_5	I_5^1	Gray clay
	Feeble confined water				I_5^2	Gray silty sand with sandy silt in-between and gas
	–				I_5^3	Gray silty clay with silty marl in-between
Hard soil of layer No. 6	–	15.00–30.0	2.00–12.60	II_1	II_1	Brown-yellow, dark green clayey soil
Sand layer of layer No. 7	I	27.00–30.0	Several m–20.00	II_2	II_2^1	Grass-yellow silty fine sand with sandy silt
					II_2^2	Gray fine silty sand
Soft soil of layer No. 8	–	25.00–50.0	Several m–35.00	I_3	II_3^1	Gray silty clay with silty and fine sand
					II_3^2	Gray silty clay and silty sand
					II_3^3	Gray, gray-green and pati-color mucky soil
Sand layer of layer No. 9	II	45.00–60.00	30.00–50.00	I_4	II_4^1	Gray fine sand with gravel
					II_4^2	Light grey fine sand with gravel

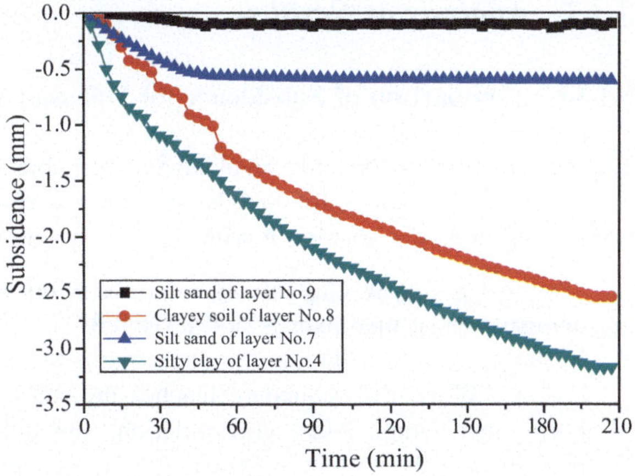

Fig. 5.8 Variations of subsidence of soil layers with time under 200 g consolidations

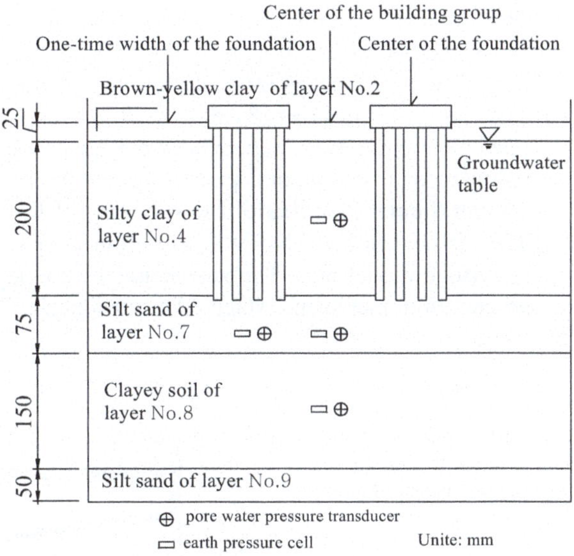

Fig. 5.9 Model section with instrumentation used

Two centrifuge model tests were conducted to study the land subsidence affected by the different building distances. The distance between two adjacent buildings in the building group was 20 and 30 m in the prototype, and the running time of the centrifuge was 6 and 8 h, respectively. The acceleration stage of the centrifuge needs 8 min, then the centrifuge reaches the stable acceleration.

5.3.3 *Analysis of Results*

5.3.3.1 Comparison of Subsidence for Different Soil Layers

The deformation of the individual layers was obtained by subtracting the subsidence recorded in an individual layer from the overall subsidence. Figure 5.10 shows variations of subsidence with time for different soil layers.

(1) The brown-yellow clay of layer No. 2 is thinner, 25 mm in thickness and deformation of this layer is not measured.
(2) In the acceleration increasing stage, the subsidence of layer No. 4 increases quickly. When the centrifuge reaches the stable acceleration, the subsidence keeps increasing. After consolidation, the subsidence value continues to increase and the soil shows rheological characteristics. The silty clay of layer No. 4 has the maximum subsidence and accounts for 43% of the total subsidence of all the soil layers.
(3) The silt sand of layer No. 7 is the bearing stratum for the pile tip and its subsidence is larger, though its thickness is 75 mm. In the acceleration stage, the subsidence reaches 60% of its total deformation.
(4) The clayey soil of layer No. 8, beneath the silt sand of layer No. 7 in which the piles were embedded, experiences the greatest compressibility resulting in the subsidence in the second place, accounting for 37% though its thickness is 50 mm thinner than that of the silty clay of layer No. 4.
(5) The silt sand of layer No. 9 is 50 mm in thickness and touches the floor of the centrifuge model box. The subsidence of this layer is small, and in the initial acceleration increasing stage, the subsidence already reaches its maximum subsidence.

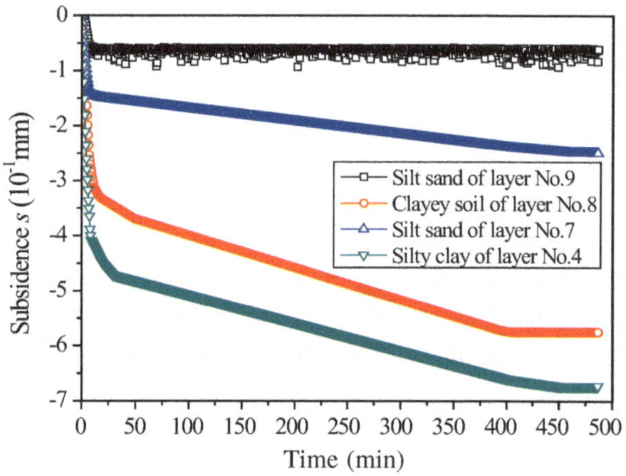

Fig. 5.10 Variations of subsidence with time for different soil layers in the test

In order to study the characteristics of every soil layer in the total subsidence stage, the subsidence of different soil layers with time is fitted with the exponential function of three-order attenuation. Figure 5.11 shows fitting curves of subsidence of different soil layers.

The exponential function of three-order attenuation is given by

$$s = A_1 e^{\frac{t}{B_1}} + A_2 e^{\frac{t}{B_2}} + A_3 e^{\frac{t}{B_3}} + s_0, \tag{5.3}$$

where A_1, A_2, A_3, B_1, B_2, and B_3 are associated with the soil properties, the thickness of soil layer, the acceleration of the centrifuge, the operating time of the centrifuge, and the building load.

This formula reflects the attenuation relationship between the subsidence and time. The velocity of subsidence gradually decreases with time and the subsidence comes to be a stable state. This formula can predict the subsidence of soil layers at any point of time.

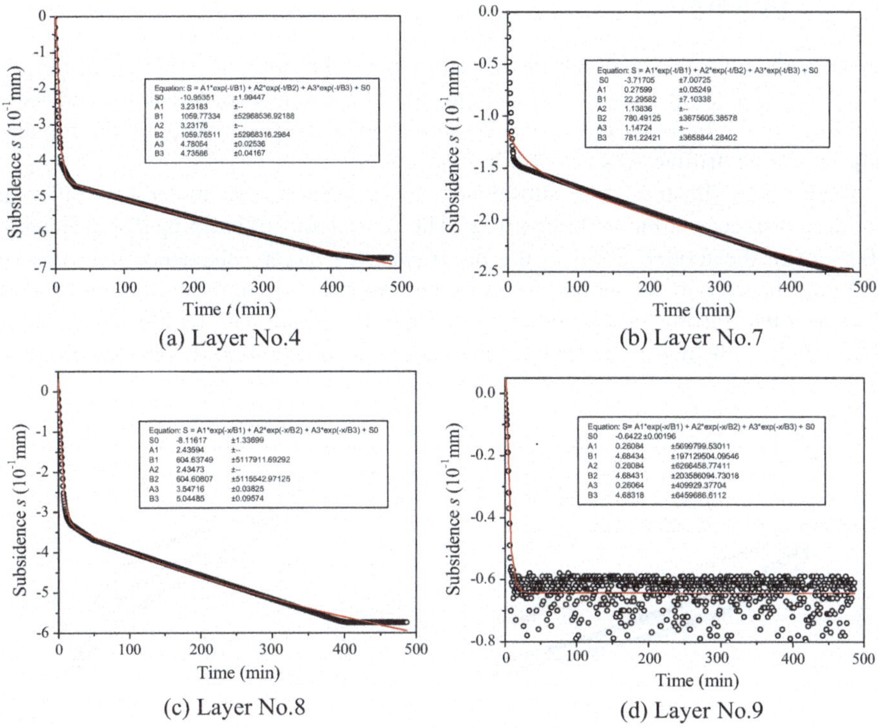

Fig. 5.11 Fitting curves of subsidence of different soil layers

5.3.3.2 Subsidence of Different Positions in the Building Group

In the acceleration increasing stage, the subsidence of different positions all increases quickly. In the stable acceleration stage, the subsidence keeps increasing and the velocity decreases. Figure 5.12 presents variations of subsidence with time at three different positions in the building group.

The central area of the high-rise building group experiences larger subsidence, so the superimposition effect of the high-rise building group in the central area is obvious. The subsidence of one time the width of the building foundation is smaller than that of the central area, which indicates that the superimposition effect is not much more obvious than that of the central area. The building itself shows the maximum subsidence. From Fig. 5.12b, the subsidence velocity of the central area is much larger than that of the building itself in the last, so the subsidence of the central area will exceed that of the building itself.

5.3.3.3 The Subsidence of the Central Area Under Different Building Distances

In order to study the land subsidence affected by the different building distances, two centrifuge model tests were conducted. The distance between two adjacent buildings in the building group was 20 and 30 m in the prototype, and the running time of the centrifuge was 6 and 8 h, respectively.

Figure 5.13 illustrates the subsidence of the central area under two different building distances in the building group. The central area of building distance being 100 mm in the model, 20 m in the prototype, has larger subsidence than that of 150 mm distance in the model, 30 m in the prototype, though the running time of the centrifuge for the smaller building distance is shorter. The building distance is smaller; the subsidence superimposition effect is more obvious. So, the building

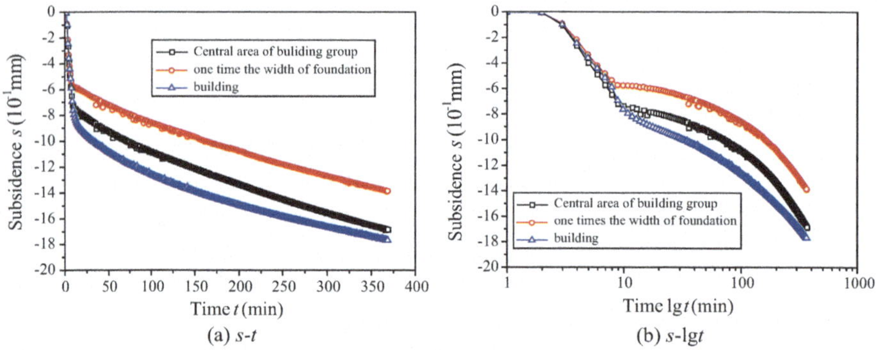

Fig. 5.12 Variations of subsidence with time at three different positions in the building group

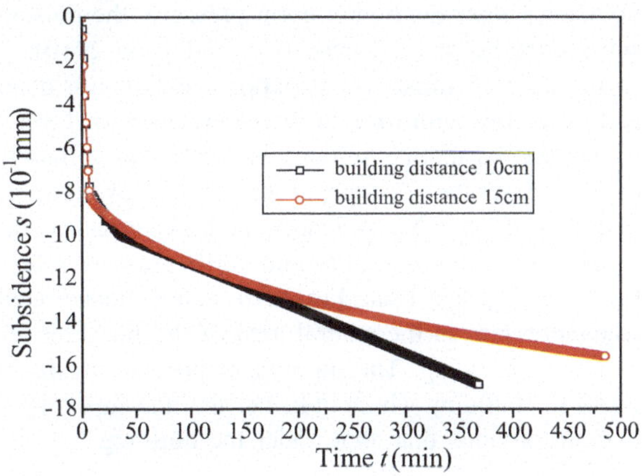

Fig. 5.13 Subsidence of the central area under two different building distances

distance among the building group can be properly increased to decrease the subsidence superimposition effect on the area.

5.4 The Subsidence with the Change of Earth Pressure and Pore Water Pressure

Figure 5.14 illustrates variations of the subsidence, the pore pressure, and the earth pressure with time in the central area of layer No. 4. In the initial stage, the earth pressure and pore pressure of the silty clay of layer No. 4 increase quickly. Then, the pore pressure begins to slowly dissipate. In the meantime, the subsidence gradually increases. The excess pore water pressure in the central area is higher which indicates the superimposition effect by the high-rise building group is

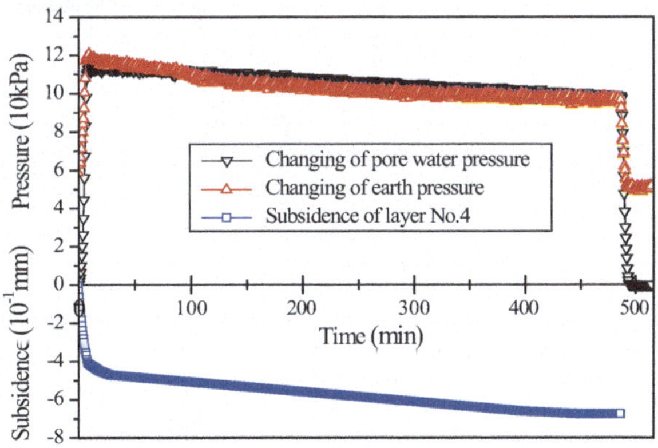

Fig. 5.14 Variations of subsidence, pore pressure, and earth pressure with time at the central area of layer No. 4

obvious. Under the higher earth pressure, the consolidation time prolongs and the subsidence keeps increasing (Cui and Tang 2010).

Figure 5.15 illustrates variations of the subsidence, the pore pressure, and the earth pressure with time in the central area of layer No. 7. In the course of subsidence, the dissipating velocity is much slower than that of layer No. 4. The buried depth of layer No. 7 is deeper than that of layer No. 4, so the earth pressure is larger than that of layer No. 4. Under the high pressure, the subsidence keeps increasing.

The pore pressure cells and earth pressure cells were embedded in the soils, shown in Fig. 5.9. From Fig. 5.16, in the course of subsidence, the trend of the pore water pressure in the central area of the building group changes the same as that under the building. The pore water pressure in the central area is higher than that under the building which indicates that the superimposition effect of the central area is more obvious than that under the building.

Fig. 5.15 Variations of subsidence, pore pressure, and earth pressure with time at the central area of layer No. 7

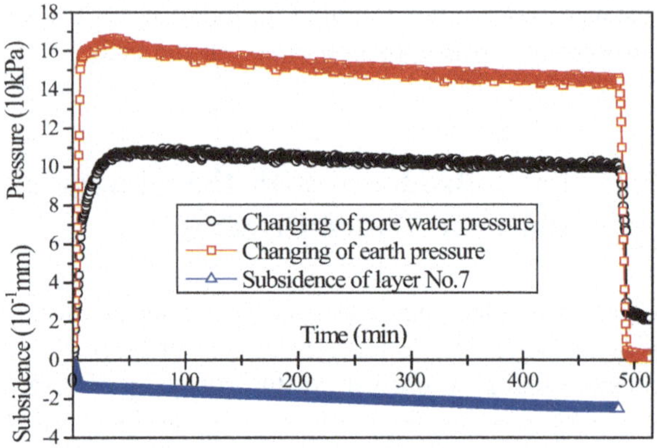

Fig. 5.16 Variations of pore water pressure with time in two different places of layer No. 7

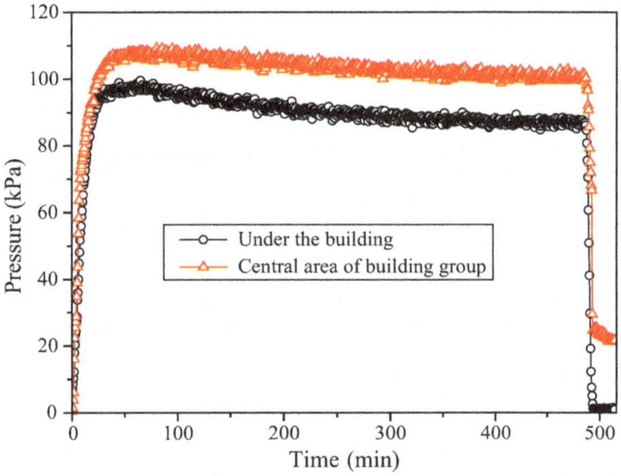

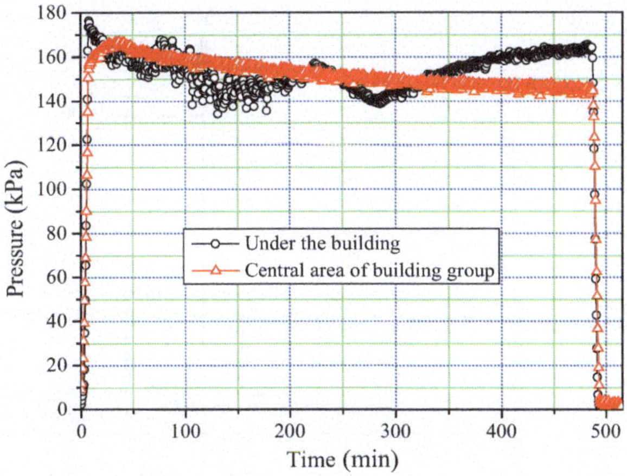

Fig. 5.17 Variations of earth pressure with time in two different places of layer No. 7

From Fig. 5.17, the earth pressure under the building fluctuates largely. The silt sand of layer No. 7 is the bearing stratum for the pile tip and the thickness of this layer is only 75 mm. In the course of subsidence, the soil around the pile tip is always disturbed by the pile. However, the earth pressure in the central area remains stable because of the fewer disturbances by the pile tip in the course of subsidence.

5.5 Three-Dimensional Numerical Simulation of the Prototype of Centrifugal Model Tests

The numerical simulation can be easily obtained for a number of useful regularities of land subsidence caused by high-rise buildings. For a clearer presentation, a series of numerical simulations were conducted to study the land subsidence caused by the interaction of four high-rise buildings and investigate the effect of different influence factors, including the building loads and the distance between buildings, etc.

5.5.1 Element Division and Material Parameters Selection

As is shown in Fig. 5.18, the soil mass of this model is divided into five layers by 160 m × 160 m × 100 m, the length of piles is 46 m and the load of the each building is 388 kPa. It is important to note that the loads and the whole model are symmetric about X and Y axes, therefore, we can build 1/4 models (Fig. 5.19) for finite-element analysis. In modeling, soil mass of this model is simulated by C3D8P element while the piles are by C3D8 element. The materials of the raft are considered as linear elastic materials and soils around the pile and below are treated as

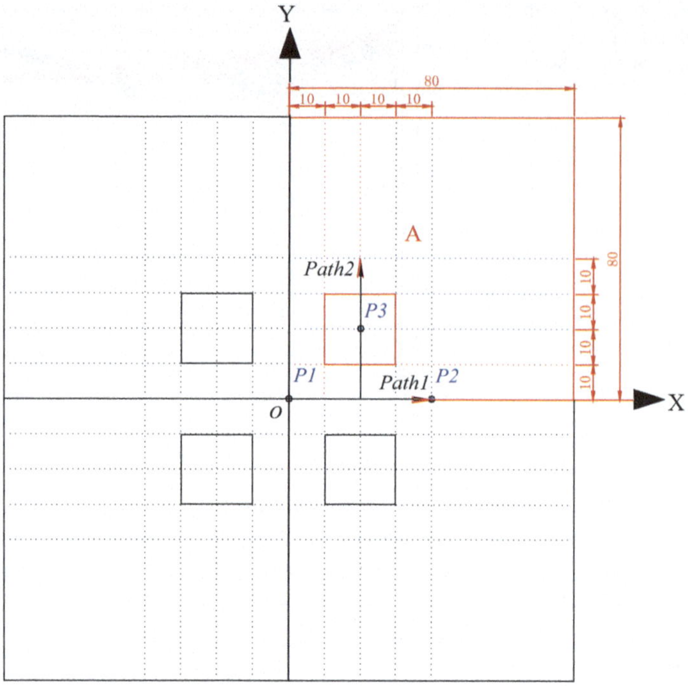

Fig. 5.18 Layout of four high-rise buildings in numerical simulation of plan

elastoplastic materials. Expressions of shear forces and shear displacement of the soils and piles are obtained as the form of penalty functions according to Mohr–Coulomb yield criterion. The specific material and load parameters of the numerical model are summarized in Tables 5.4 and 5.5.

5.5.2 Boundary Conditions

(1) Physical boundaries: displacements of lateral and bottom faces of the model do not exist.
(2) Mechanical boundaries: the top surface is free for water in soils to go while the sides and the bottom face are undrained.
(3) Initial conditions: freedom of pore water pressure is zero when the pore water is immobile. Subsidence of soil mass is steadily stated so the pore water pressure of all nodes in the model also equals zero.

Fig. 5.19 Finite-element meshes for the model

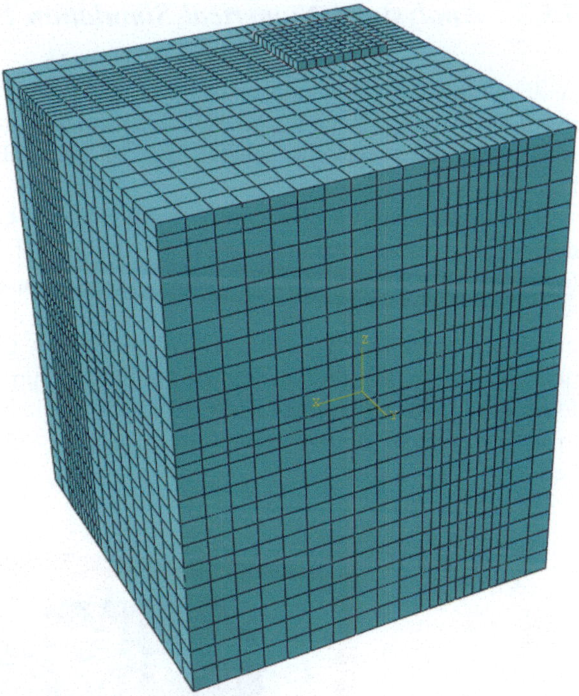

Table 5.4 Material parameters for numerical models

Soils	Elastic modulus (kPa)	Poisson's ratio (μ)	Permeability coefficient (m/d)	Gravity (KN/m^3)	Cohesion c (kPa)	Internal friction angle (°)
Pile	20,000,000	0.2	–	–	–	–
Silty clay of layer No. 4	7440	0.35	1.89	18.89	12.0	12.5
Silt sand of layer No. 7	27,150	0.25	2070	13.81	5.0	24.8
Clayey soil of layer No. 8	6390	0.38	4.67	18.88	15.8	14.5

Table 5.5 A numerical model of load parameters

Buildings	A	B	C	D
Area (m^2)	400	400	400	400
Load (kPa)	388	388	388	388

5.5.3 Analysis of Numerical Simulation Results

A 3D numerical simulation analysis of the prototype of centrifugal model test mentioned above created by ABAQUS is used to compare the results of centrifugal model test and to get the general laws of land subsidence caused by high-rise buildings (Cui et al. 2015).

To further study the subsidence of different soil layers and the displacement of the surface, subsidence nephogram of the model is obtained as is shown in Fig. 5.20, which shows clearly that a superposition of subsidence caused by four buildings exists in this simulation. The building itself and the surrounding soil undergone some degree of differential subsidence. On the other hand, three positions, including the central area of building group (P1), one time the width of foundation (P2) and building itself (P3), and two paths (Path 1, Path 2) are selected to closer analysis of the general rules of land subsidence caused by high-rise buildings, shown in Fig. 5.18.

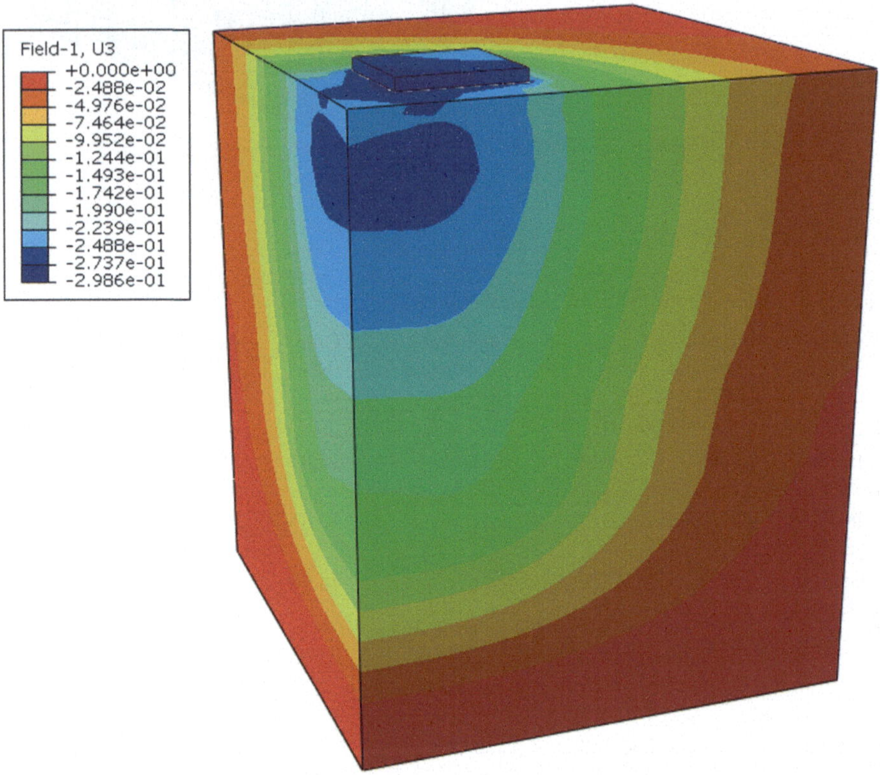

Fig. 5.20 Subsidence nephogram of the model

5.5.3.1 The Deformation of Different Soil Layers

Figure 5.21 presents variations of subsidence with time for different soil layers at the central area by numerical simulation. From Fig. 5.21, the silty clay of layer No. 4 experiences largest subsidence and the clayed soil of layer No. 8, the silt sand of layer No. 7 and the silt sand of layer No. 9 rank second, third and fourth, respectively. In addition, each soil layer will rapidly experience a larger subsidence during the loading period. The subsidence of silt sand layer of No. 9 is basically completed during the loading period. However, the silty clay of layer No. 4 and the clayed soil of layer No. 8 will have a larger consolidation subsidence after the loading period which coincides with the results of the centrifugal model tests.

5.5.3.2 Subsidence of Different Positions in the Building Group

Figure 5.22a presents variations of subsidence with time at P1, P2, and P3. The maximum subsidence occurs at P2. P1 similarly experiences larger subsidence, so the superimposition effect of the high-rise building group in the central area is obvious. On the other hand, the subsidence of P2 is smaller than that of P1, which indicates the superimposition effect is not much more obvious than that of the central area. From Fig. 5.22b, it is observed that the subsidence velocity of P1 is much larger than that of P3 in the last, so the subsidence of the central area will exceed that of the building itself. From Fig. 5.23, results of the variations of subsidence at three different positions analyzed by both the numerical simulation and the centrifugal model test are not perfect but satisfactory.

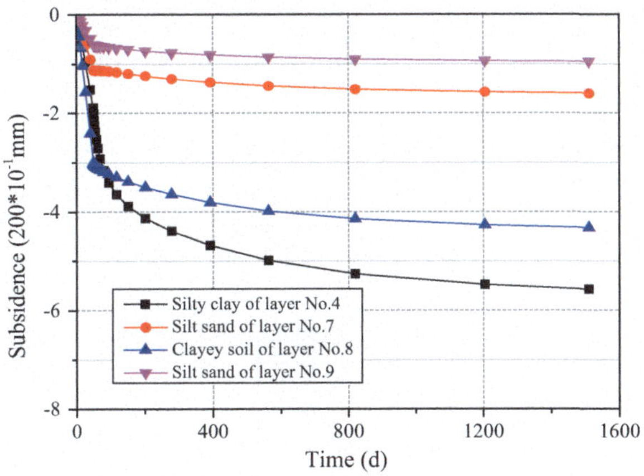

Fig. 5.21 Variations of subsidence with time for different soil layers at the central area

Fig. 5.22 Variations of subsidence with time at three different positions in the building group

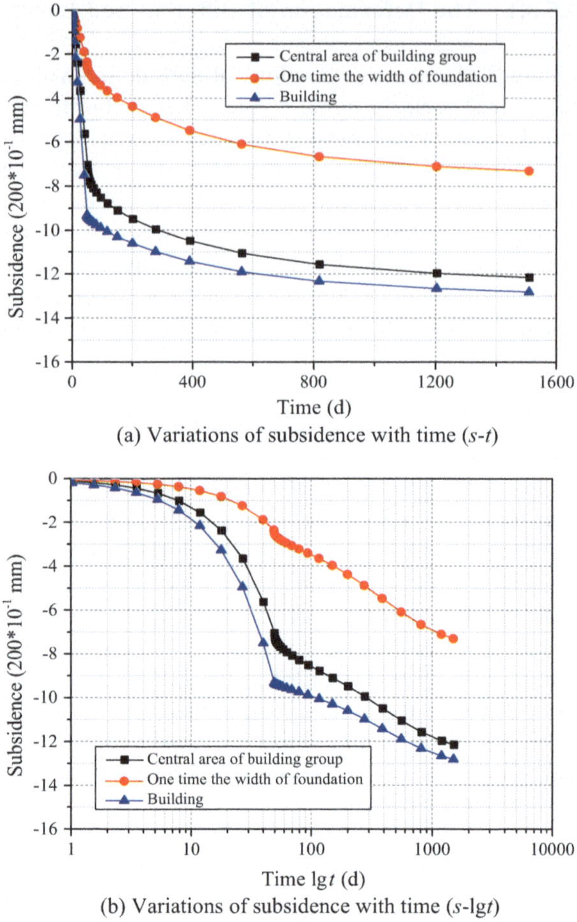

Fig. 5.23 Comparison of numerical simulation and centrifugal model tests results

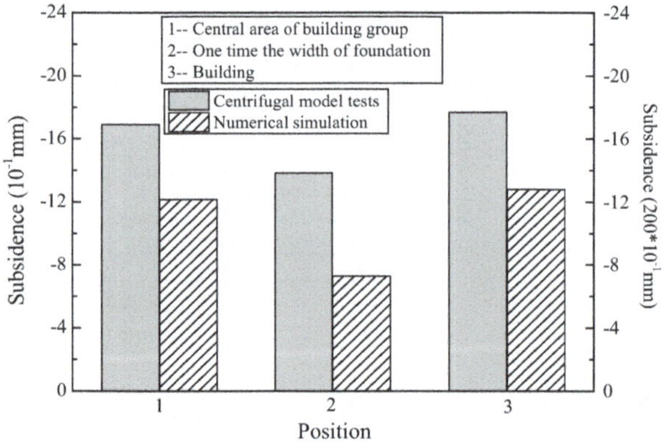

5.5.3.3 The Subsidence with the Changing of Earth Pressure and Pore Water Pressure

Figure 5.24 shows variations of subsidence, excess pore pressure and earth pressure with time in the central area of layer No. 4 and No. 7. In the course of subsidence, the dissipating velocity of layer No. 4 is much slower than that of layer No. 7. It indicates that the consolidation settlement of silty clay will be a certain degree of lag because of poor water permeability.

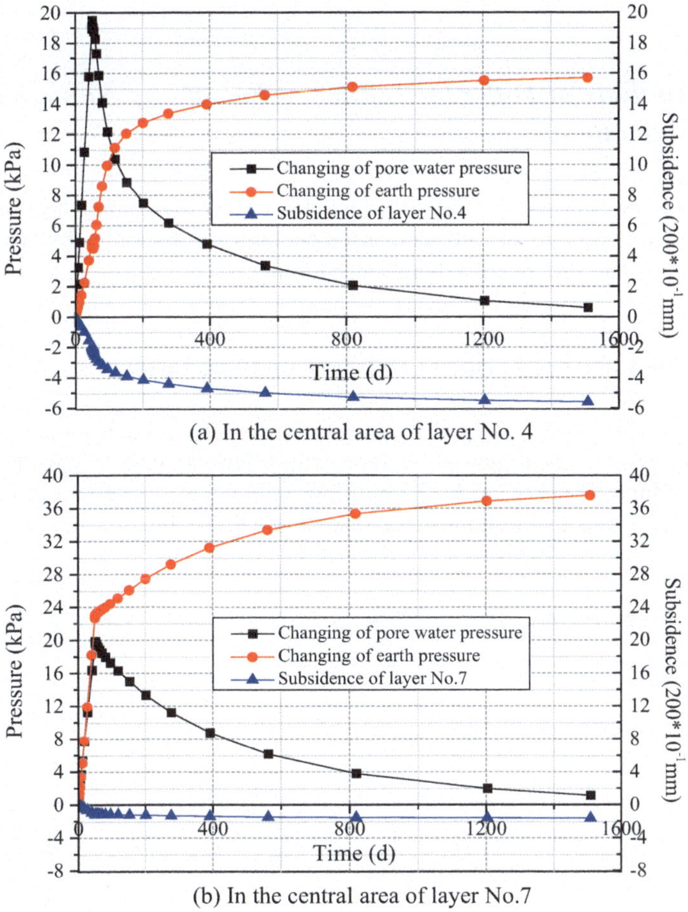

Fig. 5.24 Variations of subsidence, excess pore pressure, and earth pressure with time

5.5.3.4 Subsidence along Path 1

To further study the subsidence of different position of the surface, the subsidence along Path 1 is obtained as shown in Fig. 5.25, which shows clearly that a superposition of subsidence is caused by high-rise building group. The value of subsidence reaches to 243 mm at the central area of the buildings and the value of which are 143 mm at one time the width of foundation even though there exists an equivalent distance to the building itself in this modeling. The results indicate exactly the superposition of subsidence appears at the central area of the buildings, which matches well with the results of the centrifugal model tests.

5.5.4 Analysis of Different Influence Factors by Numerical Simulation

A series of numerical simulations are conducted to investigate the effect of different influence factors of land subsidence caused by high-rise buildings by changing the building loads and the distance between buildings.

5.5.4.1 The Influence Factor of Different Building Loads

In the interest of investigating the influence factor of land subsidence caused by different building loads. Figure 5.26a represents a few example of subsidence under different building loads (s–N) in the building foundation. As shown in Fig. 5.26a, the subsidence and the building load show the linear relationship at the initial

Fig. 5.25 Subsidence in Path 1

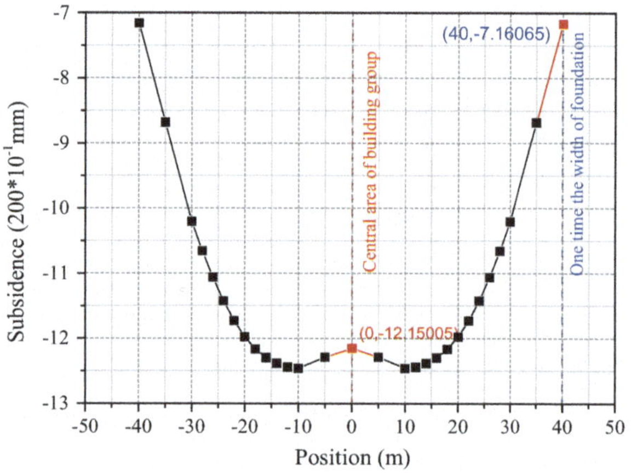

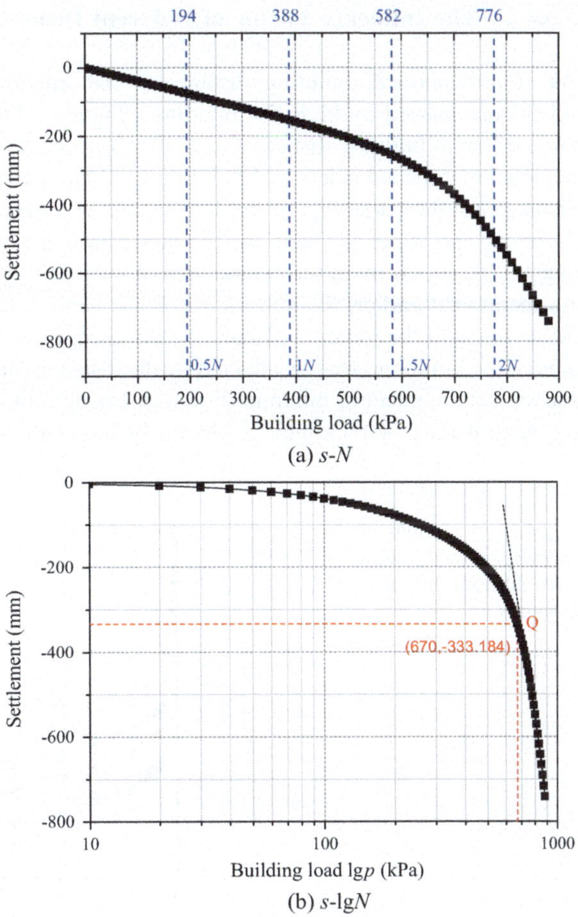

Fig. 5.26 Variations of subsidence under different building loads in the building foundation

loading stage. With the increasing load, the nonlinear relationship appears at the late loading stage. Under the conditions of building load as 0.5 N (194 kPa), 1 N (388 kPa) and 1.5 N (584 kPa), the soil layers experience the elastic deformation because of small building loads. However, large nonlinear subsidence occurs in soil layers until the building load reaches 2 N (776 kPa). This shows that the plastic deformation will occur in soil layers, causing larger land subsidence when the building load is large enough.

For a clearer presentation, Fig. 5.26b represents the curve of subsidence at different building loads (*s*–lg*N*). In order to get the ultimate bearing capacity of soils under this condition, the load value of the section (Point Q) is taken as the ultimate bearing capacity of the foundation while the slope of the curve changes sharply. As stated above, it is important to control the value of the loads in order to prevent excessive land subsidence.

5.5.4.2 The Influence Factor of Different Distances Among Buildings

As aforementioned, building distance is the important influence factors of land subsidence caused by high-rise buildings. Figure 5.27a shows the subsidence under three different building distances (20, 30, 40 m) in the building foundation. It should be noted that the subsidence of building itself decreases with the increase of building distance but the variations of the subsidence with time under three different building distances are same. In addition, the subsidence under three different building distances (20, 30, 40 m) at the central area of the building group is shown in Fig. 5.27b. As mentioned earlier, the subsidence at the central area of the building group also decreases with the increase of building distance. In fact, the subsidence at the central area of the building group is affected by the distances among buildings seriously. This shows that the effect of different distances among buildings on the land subsidence at the central area of buildings is obviously larger than that at the building itself. It

Fig. 5.27 Variations of subsidence with time under three different building distances

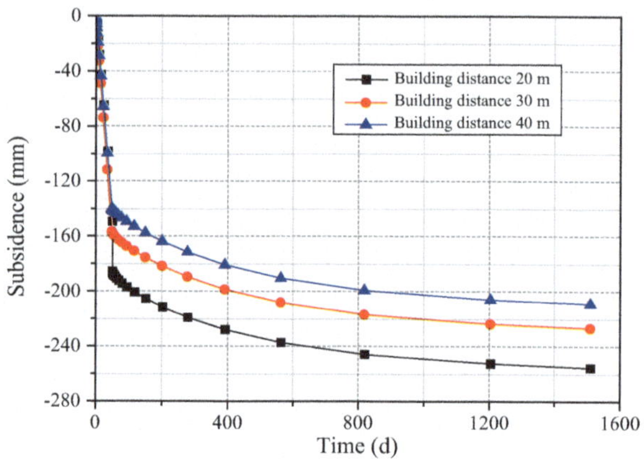

(a) In the building foundation

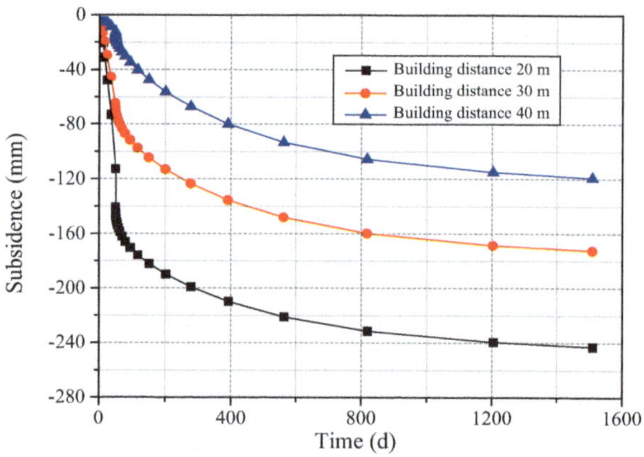

(b) At the central area of the building group

seems that we need to increase the distance of buildings as far as possible to reduce land subsidence while the loads cannot be reduced.

The subsidence under three different building distances along Path 1 is shown in Fig. 5.28a. The subsidence along Path 1 decreases with the increase of building distance obviously. On the other hand, the subsidence of one time the width of the building foundation is smaller than that of the central area, which indicates the superimposition effect is not much more obvious than that of the central area. The building distance is smaller, the subsidence superimposition effect is more obvious. Especially, the subsidence superimposition effect at the central area of buildings is obviously larger than that at other positions.

Figure 5.28b represents the subsidence under three different building distances along Path 2. This shows that the building itself has the maximum subsidence. Uneven subsidence occurs in the building foundation and its surrounding ground. Comparing three sets of data, the maximum uneven subsidence occurs at the

Fig. 5.28 Subsidence under three different building distances

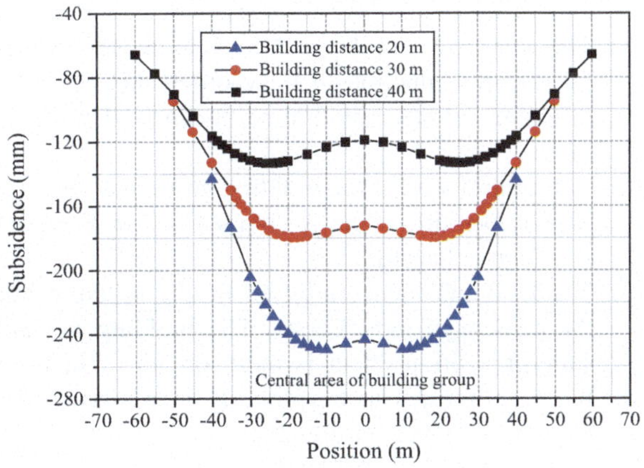

(a) Along Path1

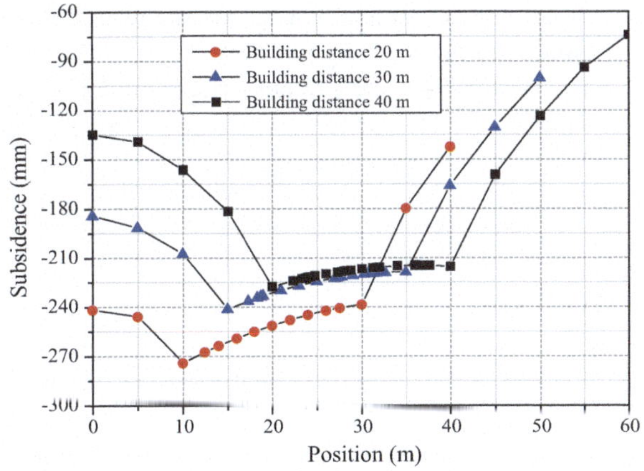

(b) Along Path2

building distance from 20 m. It indicates the superimposition effect is more obvious as the building distance decreases. In summary, the smaller distance of high-rise buildings may have obvious superimposition effect, which can cause larger land subsidence at the central area of the building group. It is feasible to increase the distance of buildings to prevent excessive land subsidence.

5.6 Conclusions

In the centrifugal model test, the model has the same stress level as the prototype and can simulate the long-term subsidence; however, the general model test (Tang and Cui 2008; Cui 2008) cannot satisfy the requirement. There are some different characteristics of the subsidence at different points and the subsidence of different soil layers between the centrifuge model test and the general one. But the general trends of land subsidence are in good agreement with each other.

(1) The silty clay of layer No. 4 experiences the maximum subsidence. The exponential function of three-order attenuation can well fit the subsidence of different soil layers with time. It can predict the subsidence of soil layers at any time.
(2) The central area of the building group has larger subsidence and the subsidence superimposition effect is obvious. It can exceed the allowance and cause land subsidence hazard.
(3) The land subsidence affected by the different building distances is studied by the centrifuge model tests. The building distance is smaller; the subsidence superimposition effect is more obvious. So the building distance among the building group can be properly increased to decrease the subsidence superimposition effect on the area.
(4) In the centrifuge model tests, the larger excess pore water pressure in the bearing stratum for the pile tip of the central area also shows that the stress superimposition effect is larger at the central area. The earth pressure under the buildings fluctuates by the disturbance of the pile tip.
(5) The results from the 3D numerical simulation match well with the centrifugal model test which can be used to analyze and to predict the land subsidence of high-rise buildings.

References

Cui ZD (2008) Study on the land subsidence caused by the dense high-rise building group in the soft soil area. Ph.D. thesis, Tongji University

Cui ZD (2012) Bearing capacity of single pile and in-flight T-bar penetration for centrifuge modeling of land subsidence caused by the interaction of high-rise buildings. Bull Eng Geol Environ 71:579–586

Cui ZD, Tang YQ (2010) Land subsidence and pore structure of soils caused by the high-rise building group through centrifuge model test. Eng Geol 113:44–52

Cui ZD, Tang YQ, Yan XX (2010a) Centrifuge modeling of land subsidence caused by the high-rise building group in the soft soil area. Environ Earth Sci 59:1819–1826

Cui ZD, Yang JQ, Yuan L (2015) Land subsidence caused by the interaction of high-rise buildings in soft soil areas. Nat Hazards 79:1199–1217

Cui ZD, Tang YQ, Yan XX, Yan CL, Wang HM, Wang JX (2010b) Evalutaion of the geology-environmental capacity of buildings based on the ANFIS model of the floor area ratio. Bull Eng Geol Environ 69:111–118

Geddes JD (1966) Stresses in foundation soils due to vertical subsidence load. Geotechnique 16 (3):231–255

Powrie W, Daly MP (2007) Centrifuge modeling of embedded retaining walls with stabilising bases. Geotechnique 57(6):485–497

Tang YQ, Cui ZD (2008) Model test study of land subsidence caused by the high-rise building group. Bull Eng Geol Env 67(2):173–179

Chapter 6
Microstructures of Different Soil Layers Before and After Centrifuge Modeling of Land Subsidence Caused by High-Rise Building Group

6.1 Introduction

The engineering characteristics of soils are controlled by the state of pore structure of soils to a great extent. SEM is frequently employed to analyze the microstructure of soils in an attempt to assess their environmental and mechanical behavior (Mitchell 1993). Cetin et al. (2007) studied the microstructure of a cohesive soil during compaction by SEM test. Shi et al. (1999) presented the preparation of soil specimens for SEM test. Yamamuro and Wood (2004) investigate the effect of depositional method on the microscopic grain structures of sand containing non-plastic silt by SEM.

The macroscopic characteristics of soils such as the permeability, the compressibility, and the strength are controlled by the parameters of soil microstructure such as the size distribution of particles and pores, the contact ways of particles, and the morphology of pores to a great extent (Hong et al. 2006). Every continuum model depending on the certain homogenization was difficult to explain the complexities of the soil microstructures (Tan 1983). The early researches on the soil microstructures were intended to explain the soil properties qualitatively. With the development of test technology, quantitative researches on the soil microstructures developed rapidly. The X-ray diffraction (XRD) was conducted to determine the mineral composition of the soil (Lin and Cerato 2012). The mercury intrusion porosimetry (MIP) tests were conducted to study the pore structure of the soil (Cui et al. 2010; Zhang and Li 2010). Although XRD and MIP were reliable and repeatable, they were limit to quantitatively study the soil microstructure, and the test results were only suitable for the certain conditions. The parameters of the soil microstructure such as the diameter, the shape factor, and the anisotropy rate of particles or pores were obtained by the X computed tomography (X-CT) or the scanning electron microscopy (SEM) combined with image processing (Sleutel et al. 2008; Shi 1996), which can rebuild the three-dimensional microstructure of the soil (Wang et al. 2004a, b). By this method, many results can be obtained,

Z.-D. Cui, *Land Subsidence Induced by the Engineering-Environmental Effect*,
https://doi.org/10.1007/978-981-10-8040-1_6

but the accuracy was less. It was difficult to accurately determine what grayscale value represented the particle or the pore in a certain grayscale image. The porosity obtained from the two-dimensional image processing was affected by the selected threshold greatly (Wang et al. 2008) while vertical and horizontal scale of the image should be adjusted based on the experience for the three-dimensional processing (Wang et al. 2011). The threshold range [60, 100] was suggested when studying the pore structure quantitatively and the range [150, 220] for studying the particle morphology (Wang et al. 2004a, b). However, the suggested threshold range was not accurate enough. The soil microstructures were quantitatively studied by some researchers through the MIP tests, accompanying with the X-CT or SEM tests for qualitative analysis of images to ensure the validity (Cui and Tang 2011). The obtained conclusions were more available, but the X-CT or SEM tests for quantitatively studying were not functioned enough for less accuracy of the image processing.

Mercury intrusion porosimetry (MIP) has been routinely used to evaluate the pore size distribution of powdered and bulk materials with open and interconnected pore structures. Recent applications of mercury intrusion porosimetry in geotechnical engineering include the following: the effect of soil microstructure on the compressibility of natural Singapore marine (Low et al. 2008); the compressibility effect in evaluating the pore size distribution of kaolin clay (Penumadu and Dean 2000); the differences between the pore size distribution in laboratory and field compacted soil (Prapaharan et al. 1991); the feasibility of correlating permeability with the pore size distribution of clays (Lapierre et al. 1990); the effect of air drying and critical point drying on the porosity of clay soils (De Kimpe 1984); permeability as a function of the pore size distribution of silty clay (Garcia-Benaochea et al. 1979); and the pore size distribution in clays (Diamond 1970).

This chapter studied the microstructure of each soil layer under the building loads in the centrifuge model by SEM and MIP, for qualitative analysis and quantitative analysis, respectively. Before and after the centrifuge model test conducted in Chap. 5, samples of each soil layer were prepared for studying the microstructure of soils by SEM test and MIP test.

6.2 The Scanning Electron Microscopy Test

6.2.1 Soil Preparation

Soil samples of silty clay of layer No. 4, silt sand of layer No. 7, clayey soil of layer No. 8, and silt sand of layer No. 9 were prepared before and after the centrifuge model test, except brown-yellow clay of layer No. 2, 25 mm in thickness. The preparation process of soil samples is as follows:

(1) The soil sample should be broken by hand, not by mechanical cutting tools. Breaking the sample well is not easy. If there is smooth internal surface of the

sample, it is easy to be broken along the smooth surface, and the observed microstructure of sample is the feature of this smooth surface. So attention should be paid to choose the part on behalf of the soil microstructure in the process of sample preparation.

(2) Break the soil samples into certain sizes along the observing direction and then choose smooth sections of the samples with a magnifying glass.
(3) Then, put the samples in a drying case at a certain temperature less than 100 °C for a certain time.
(4) After drying, put the samples into a spraying machine and spray with conductive material being gold or carbon-gold.

6.2.2 *Qualitative Analysis of Test Results*

The amplifications of the SEM test are 3000× and 500× for silty clay of layer No. 4 and clayey soil of layer No. 8, and silt sand of layer No. 7 and layer No. 9, respectively. Figure 6.1 shows the microstructure of each soil layer by SEM. The microstructure of each soil layer presents states of face–face contact, edge–edge contact, and face–edge contact.

For silty clay of layer No. 4 and clayey soil of layer No. 8, before the centrifuge model test, particles show good arrangement and take on the shape of flake. After the centrifuge model test, particles rearrange tightly and the consolidation was conducted. The pore size becomes smaller.

Because silt sand of layer No. 7 is the bearing stratum of the pile tip and layer No. 9 lies at the bottom of strongbox and touch the floor of strongbox, soil particles are damaged a little after the test. The pore size also becomes smaller.

6.2.3 *Quantitative Analysis of Test Results*

The amplifications of the SEM test were 500×, 1000×, 3000×, 5000×, and 10,000× for the silty clay of layer No. 4 and the clayey soil of layer No. 8, and 100×, 500×, 1000×, 3000×, 5000×, and 10,000× for the silt sand of layer No. 7 and the silt sand of layer No. 9. An appropriate amplification of the image should be selected to study the microstructure of soils quantitatively with different particle sizes. Figure 6.1 shows the SEM images with appropriate amplifications, 3000× for the silty clay of layer No. 4 and the clayey soil of layer No. 8 and 500× for the silt sand of layer No. 7 and layer No. 9, respectively. Section 6.2.2 qualitatively analyzed the microstructures of different soil layers caused by the high-rise building group in Shanghai. This chapter mainly studies the changes of soil microstructure characteristics quantitatively before and after the centrifuge model test.

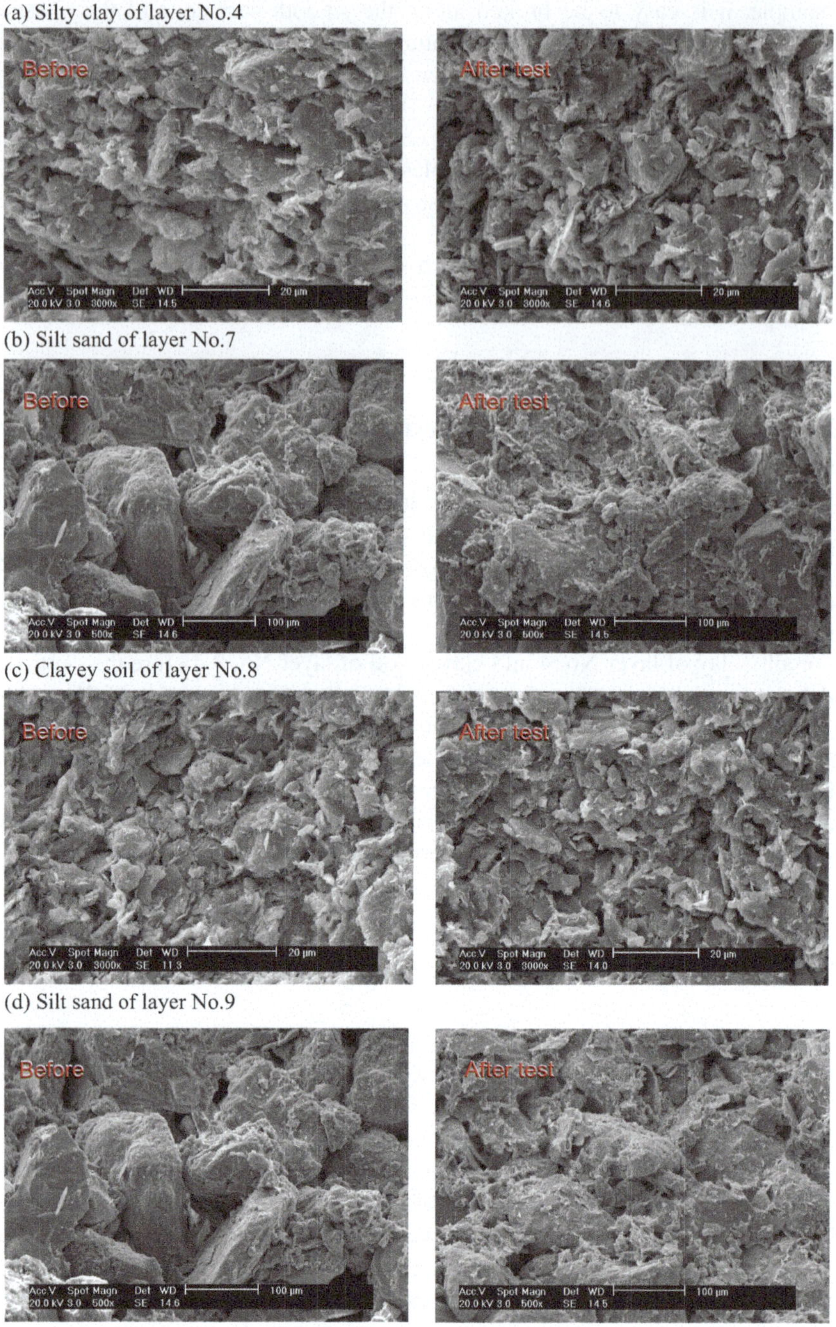

Fig. 6.1 Microstructure of each soil layer before and after the centrifuge model test

6.2.3.1 The Calculation of Void Ratio from SEM Images

The image processing was conducted to obtain characteristics of the pore structures, including reading the image, converting the format of the image, and calculating the parameters of the pore structures. The void ratio calculated was controlled by the selected threshold for the binary of the images. The threshold was normalized, noted as T_n, having the value from 0 to 1 with the interval of 0.1 for images with different amplifications of different soil layers before the centrifuge model test. If the normalized grayscale value of a certain pixel was smaller than or equal to that of the selected T_n, it was identified as the pore; otherwise, it was identified as the particle. The area of particles can be calculated by counting the white pixels. The area of the particle was named as the filling rate n', and the following function can be obtained:

$$n + n' = 1, \tag{6.1}$$

where n is the rate of the pores, and n' is the filling rate of the particles.

Take the 3000× SEM image of the silty clay of layer No. 4 as an example. Figure 6.2 shows the distribution of the grayscale, and Fig. 6.3 shows the pore structure with T_n being 0.2, 0.3, and 0.4, respectively. From Fig. 6.2, the grayscale values of 3000× SEM image of the silty clay of layer No. 4 are mainly varying from 50 to 100. From Fig. 6.3, the larger value of T_n, the more pixels will be identified as the pore. To the partition of the foreground and background for a certain image, the Otsu method is available. It is assumed that T is the dividing threshold between the foreground and the background, ω_0, ω_1 being the ratio of the

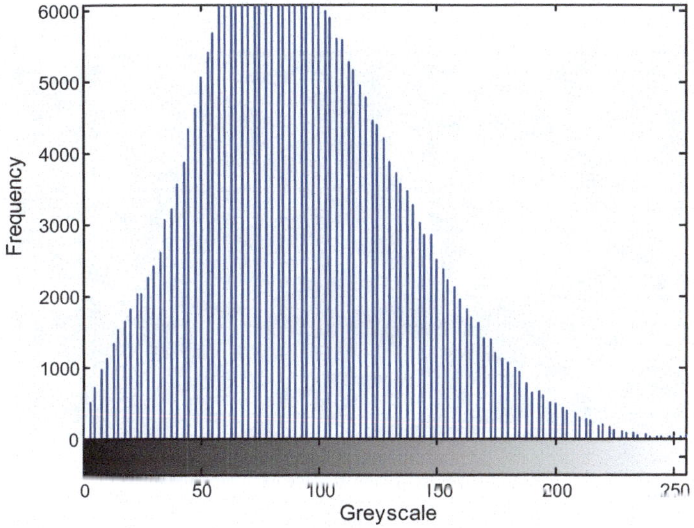

Fig. 6.2 Grayscale distribution of 3000× SEM image of silty clay of layer No. 4

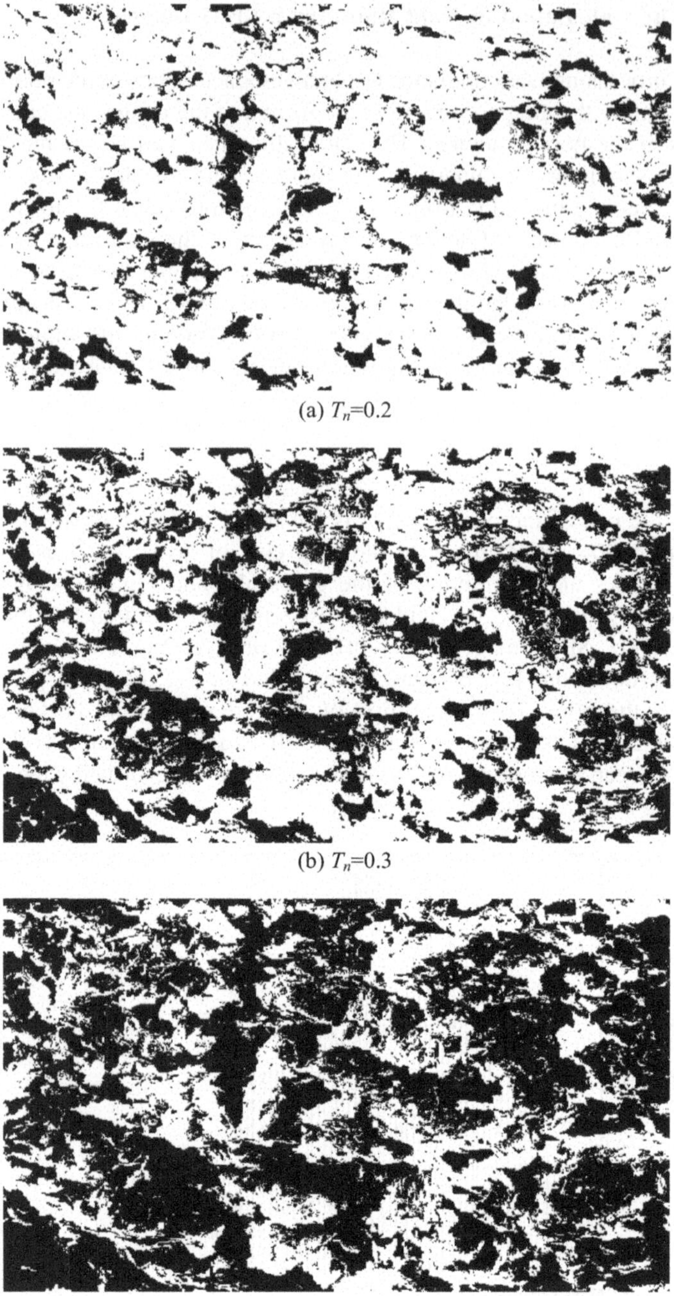

(a) T_n=0.2

(b) T_n=0.3

(c) T_n=0.4

Fig. 6.3 Pore structure of silty clay of layer No. 4

pixels in the foreground and in the background to all the pixels, respectively, u_0, u_1 being the mean grayscale value of the pixels in the foreground and in the background, respectively, the mean grayscale value of the whole image is $u_T = \omega_0 \times u_0 + \omega_1 \times u_1$. When the variance $\sigma^2 = \omega_0(u_0 - u_T) + \omega_1(u_0 - u_T)^2$ reaches the maximum, T is the best dividing threshold. Variations of n' of different soil layers with different T_n calculated by the Otsu method are summarized in Table 6.1, compared with the real filling rate.

Table 6.1 Calculated n' with different T_n before consolidation

T_n	n'					Real n'
	500×	1000×	3000×	5000×	10,000×	
(a) *Silty clay of layer No. 4*						
0	0.913	0.9996	0.998	1	0.9774	0.482
0.1	0.7835	0.9812	0.9545	0.9735	0.9185	0.482
0.2	0.6089	0.9042	0.8426	0.8652	0.7854	0.482
0.3	0.4321	0.7376	0.6205	0.6361	0.5475	0.482
0.4	0.2874	0.5119	0.3921	0.3784	0.3211	0.482
0.5	0.1809	0.3026	0.2153	0.1924	0.1716	0.482
0.6	0.1022	0.1388	0.0926	0.0724	0.0666	0.482
0.7	0.0567	0.0581	0.035	0.0223	0.0182	0.482
0.8	0.0282	0.0207	0.01	0.0051	0.0039	0.482
0.9	0.0117	0.0059	0.0018	0.00065	0.0004	0.482
1	0	0	0	0	0	0.482
Otsu	0.3695	0.4232	0.4133	0.4018	0.4009	0.482
Otsu error (%)	−23.3	−12.2	−14.25	−16.64	−16.83	–
T_n	n'					Real n'
	500×	1000×	3000×	5000×	10,000×	
(b) *Clayey soil of layer No. 8*						
0	0.9902	0.9843	0.9963	0.9964	0.9929	0.5
0.1	0.932	0.9123	0.9529	0.9515	0.938	0.5
0.2	0.7812	0.7581	0.8325	0.8301	0.7811	0.5
0.3	0.5503	0.5493	0.627	0.6125	0.5326	0.5
0.4	0.3441	0.3594	0.4151	0.389	0.3111	0.5
0.5	0.1978	0.2174	0.2465	0.2184	0.159	0.5
0.6	0.0936	0.1108	0.1194	0.0948	0.0685	0.5
0.7	0.0396	0.051	0.0519	0.0392	0.0266	0.5
0.8	0.0131	0.109	0.0179	0.0136	0.0084	0.5
0.9	0.0029	0.0051	0.0046	0.004	0.0022	0.5
1	0	0	0	0	0	0.5
Otsu	0.381	0.3934	0.4151	0.389	0.3912	0.5
Otsu error (%)	−23.8	−21.3	−17.0	−22.2	−27.2	–

(continued)

Table 6.1 continued

T_n	n'						Real n'
	100×	500×	1000×	3000×	5000×	10,000×	
(c) *Silt sand of layer No. 7 (No. 9)*							
0	0.9822	0.9878	0.9954	0.9904	0.9962	0.9998	0.467
0.1	0.8639	0.934	0.9447	0.9338	0.9557	0.9801	0.467
0.2	0.6435	0.7827	0.7878	0.8086	0.7925	0.8841	0.467
0.3	0.4098	0.5042	0.5628	0.5937	0.5487	0.672	0.467
0.4	0.2455	0.295	0.344	0.3955	0.303	0.4265	0.467
0.5	0.1406	0.1666	0.1876	0.2498	0.123	0.2552	0.467
0.6	0.0727	0.083	0.0811	0.1242	0.0454	0.1316	0.467
0.7	0.0384	0.0378	0.0309	0.0513	0.0194	0.0599	0.467
0.8	0.0186	0.0154	0.0101	0.0146	0.0009	0.0222	0.467
0.9	0.008	0.0084	0.0039	0.0034	0.0037	0.0057	0.467
1	0	0	0	0	0	0	0.467
Otsu	0.3025	0.3116	0.383	0.3955	0.4491	0.349	0.467
Otsu error (%)	−35.2	−33.3	−18.0	−15.3	−3.8	−25.3	–

The minimum and maximum errors of n' by the Otsu method for the silty clay of layer No. 4, the clayey soil of layer No. 8, and the silt sand of layer No. 7 (No. 9) are −12.2, −17.0, −3.8 and −23.3, −27.2, −35.2%, respectively. It is not accurate enough to calculate the filling rate by the Otsu method. However, the real filling rate can be estimated by adding 15% to the calculated filling rate of the particle with the T_n decided by the Otsu method. To calculate the filling rate from SEM images, the adopted T_n should be between the range of [0.3, 0.4], that is, the corresponding grayscale value should be [75, 100]. However, Fig. 6.2 shows that the grayscale values of most pixels are between [50, 100]. So, the range [75, 100] is not accurate enough. For example, when T_n equals 0.3, the minimum and maximum errors of the calculated n' for the silty clay of layer No. 4, the clayey soil of layer No. 8, and the silt sand of layer No. 7 (No. 9) are −10.2, 9.9, 8.0 and 52.2, 25.4, 43.9%, respectively. When T_n equals 0.4, the minimum and maximum errors of the calculated n' for the silty clay of layer No. 4, the clayey soil of layer No. 8, and the silt sand of layer No. 7 (No. 9) are 6.2, −17.0, −8.67 and −40.4, 37.8, −47.43%, respectively. So the range needs to be narrowed.

The corresponding T_n to the real filling rate can be back-calculated by the fitting function between n' and T_n. The fitting method affects the results to a great extent. The polynomial fitting method and the Gaussian fitting method were used. Figure 6.4 shows the comparison of the fourth degree polynomial fitting with the

Fig. 6.4 Comparison of fourth degree polynomial and Gaussian fitting

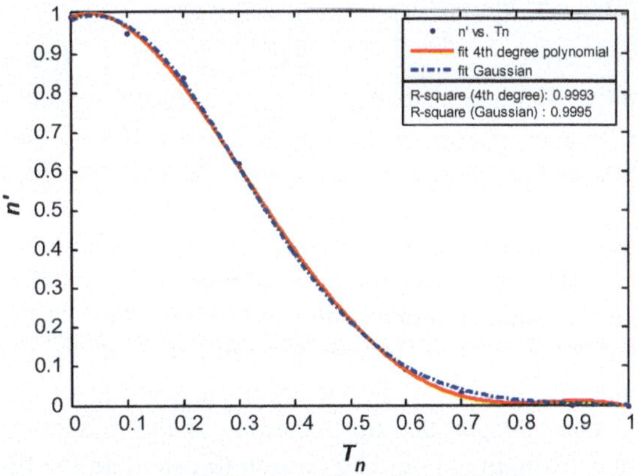

(a) Silty clay of layer No. 4 (3000×)

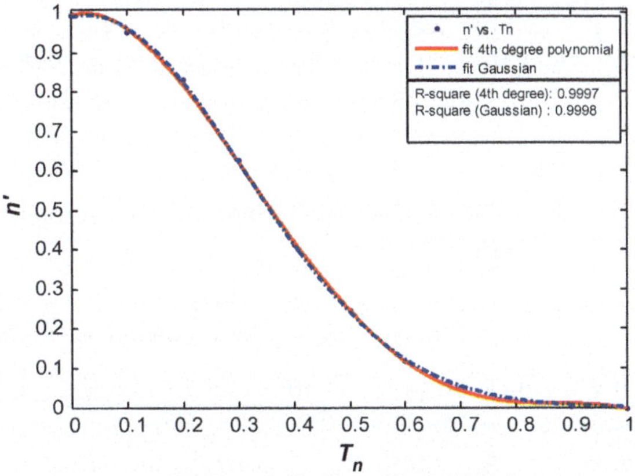

(b) Clayey soil of layer No. 8 (3000×)

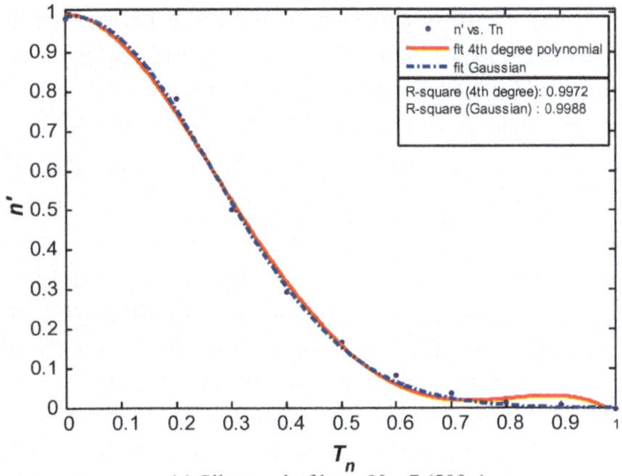

(c) Silty sand of layer No. 7 (500×)

Table 6.2 Relationship between n' and T_n

Soil layers	100×	500×	1000×	3000×	5000×	10,000×
Silty clay of layer No. 4	–	0.2730	0.4069	0.3578	0.3557	0.3267
Clayey soil of layer No. 8	–	0.3257	0.3986	0.3497	0.3483	0.3188
Silt sand of layer No. 7	0.2809	0.3248	0.3415	0.3652	0.3266	0.3877

Gaussian fitting for the silty clay of layer No. 4 (3000×), the clayey soil of layer No. 8 (3000×), and the silt sand of layer No. 7 (500×). The relationship between n' and T_n is in good agreement with that of the Gaussian fitting. The value of T_n back-calculated by the Gaussian function is summarized in Table 6.2.

The appropriate amplifications of the images which can fully represent the soil microstructure should be chosen to calculate the filling rate of the particles with the SEM images. 3000× or 5000× is suitable for the silty clay of layer No. 4 and the clayey soil of layer No. 8 and 500× or 1000× is suitable for the silt sand of layer No. 7 and No. 9. The selected T_n should be the value between the range [0.34, 0.36], which corresponds to the value between the range [85, 90] for the real grayscale.

6.2.3.2 Microscopic Mechanism of Soil Consolidation and Land Subsidence

(1) The sizes and morphological characteristics changes of pores

How to pick up pores from a certain image is the key to obtain the quantitative parameters of the microstructures. Here are two methods: one is the back calculation of the threshold, binarizing the images with the calculated threshold according to the void ratio obtained from laboratory test; the other is picking up pores depending on the similar value of the grayscale, which is treating the pixels with the grayscale value similar to the apparent pores distinguished by eyes as the pores. The SEM images of the silty clay of layer No. 4 (3000×) were taken as an example to compare the two methods. According to Table 6.2, the corresponding T_n was 0.3578. Figure 6.4 shows the pore structure extracted by the two methods.

Figure 6.5 shows that the filling rate is near to the real filling rate, but some particles are mistakenly identified as pores. The reason was that the SEM images can only illustrate the inter-aggregate pores but little intra-aggregate pores. The grayscale values of different soils in the same height are also different. So, only when the void ratio calculated with the first method is less than the real void ratio, it can represent the real soil microstructure, and the difference depends on the intra-aggregate void ratio, which is often estimated by the experience. So, the binarization of the SEM images can only be used to estimate the macroscopic void ratio instead of calculating the soil microstructure characteristics accurately. The inter-aggregate pores can be extracted directly with the second method, and the quantitative analysis can be undertaken. However, what value of the grayscale

Fig. 6.5 Pore structure of silty clay of layer No. 4 (3000×)

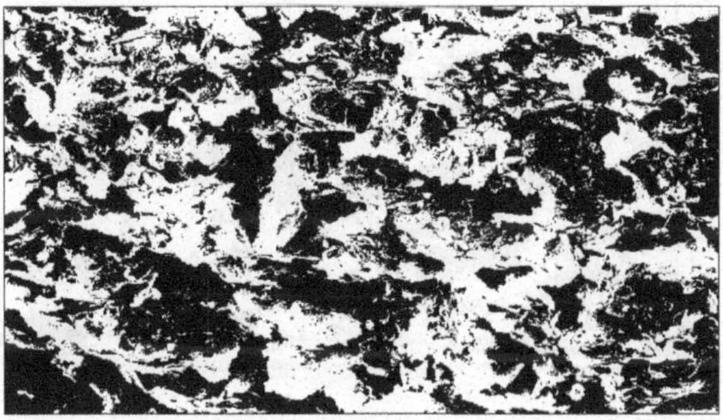

(a) Method of back calculation of threshold (n' =0.479)

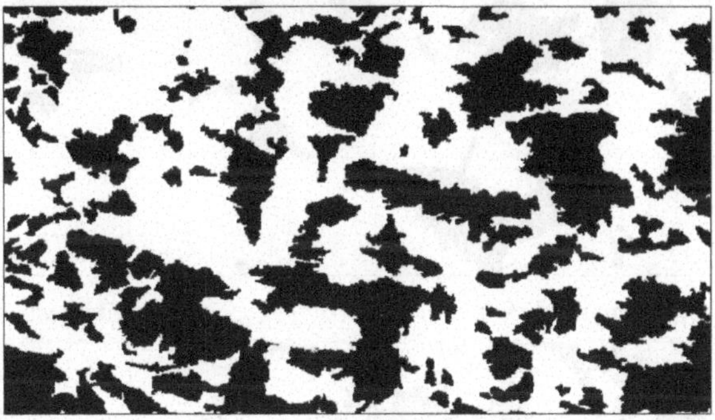

(b) Method of picking up pores depending on similar grayscale value (n'=0.571)

representing the particle or the pore in a certain image did not have the unified conclusion yet, so the result included the subjective error more or less. One method to reduce the error was to magnify the images to a certain extent when extracting the pores. This chapter used the method of extracting the pores depending on the similar grayscale value. Figure 6.6 shows the pore structures of the silty clay of layer No. 4, the clayey soil of layer No. 8, the silt sand of layer No. 7, and the silt sand of layer No. 9 before and after the centrifuge model test. After the centrifuge model test, the filling rate n' of the silty clay of layer No. 4, the clayey soil of layer No. 8, the silt sand of layer No. 7, and the silt sand of layer No. 9 increased from 0.571, 0.790, 0.731, and 0.790 to 0.691, 0.900, 0.813, and 0.844, respectively. This indicated that each soil layer was compacted to different extents. With the pore structure images, the pore microscopic characteristics were calculated, including the numbers of pores N, the area A of each pore, the equivalent diameter D_E of each pore, the perimeter P of each pore, and the ellipse which has the same standard secondary moment with each pore. So the average area $\bar{A}$, the average equivalent diameter $\bar{D}_E$, the average eccentricity $\bar{e}$, the average circularity $\bar{R}$, and the average shape factor F were obtained. It was noted that in the image, the unit of the length was the pixel. For analyzing the characteristics of the pores quantitatively, the

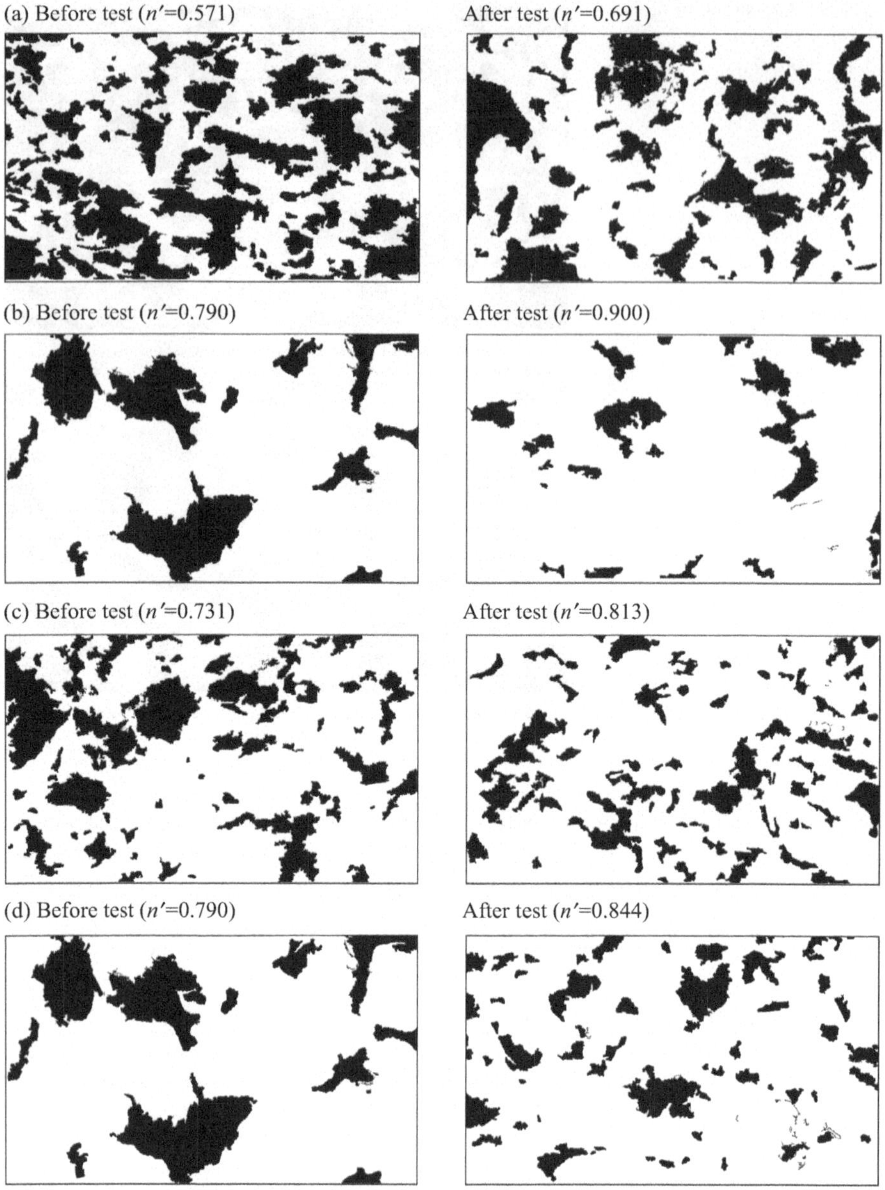

Fig. 6.6 Pore structure of each soil layer before and after the centrifuge model test

length with the unit pixel should be transformed to the real length with the unit μm. To the 3000× and 500× amplification images, one pixel represents 0.1314 and 0.7876 μm, respectively. The pore morphological characteristics of each soil layer before and after the centrifuge model test are summarized in Table 6.3.

Table 6.3 Pore morphological characteristics of each soil layer before and after the centrifuge model test

Characteristics	Silty clay of layer No. 4		Clayey soil of layer No. 8		Silt sand of layer No. 7		Silt sand of layer No. 9	
	Before	After	Before	After	Before	After	Before	After
n	63	65	65	73	14	24	14	68
$\bar{A}/\mu m^2$	35.11	24.80	23.93	13.39	2780	770.0	2780	424.0
$\bar{D}_E/\mu m$	5.2440	4.1435	4.2508	3.311	47.3830	25.0578	47.3830	18.8818
$\bar{e}$	0.7882	0.8183	0.8163	0.8447	0.8470	0.8514	0.8470	0.8131
$\bar{R}$	0.4419	0.4134	0.4265	0.3788	0.3837	0.3663	0.3837	0.4492
$\bar{F}$	0.5770	0.5934	0.6019	0.6247	0.5184	0.5325	0.5184	0.6586
$\overline{a_E}/\mu m$	8.960	6.871	7.093	5.927	77.39	45.37	77.39	23.97

Note n is the pore number in each image

$\bar{A} = \frac{\sum_{i=1}^{n} A_i}{n}$; $\bar{D}_E = \frac{\sum_{i=1}^{n} D_{Ei}}{n}$; $\bar{e} = \frac{\sum_{i=1}^{n} e_i}{n}$; $\bar{R} = \frac{\sum_{i=1}^{n} R_i}{n}$; $R = \frac{A}{A'}$, where A' is the circumcircle area of a certain pore.

$\bar{F} = \frac{\sum_{i=1}^{n} F_i}{n}$; $F_i = \frac{P_i}{P_i'}$, where P_i is the perimeter of the circle which has the same area with a certain pore; P_i' is the perimeter of a certain pore.

$\overline{a_E} = \frac{\sum_{i=1}^{n} a_{Ei}}{n}$, where a_{Ei} is the long axis length of the ellipse which has the same standard secondary moment with a certain pore.

Before the centrifuge model test, the edges and corners of the silt sand of layer No. 7 and layer No. 9 were clear, and the inter-aggregate pores were big. Because the silt sand of layer No. 7 was the bearing stratum of the pile tip and the layer No. 9 lied at the bottom of the strongbox, soil particles were damaged a little during the centrifuge model test and produced the finer particles. The pore size became smaller and the pore numbers in the unit area increased. The grayscale value of the finer particles was bigger, the brightness was higher, which affected the extraction of pores, and the error occurred. The subsidence of silt sand of layer No. 7 and No. 9 accounted for 20% of the total land subsidence (Cui et al. 2010). This chapter mainly studied the microscopic mechanism of the silty clay of layer No. 4 and the clayey soil of layer No. 8, which experienced 80% of the total land subsidence.

The parameters describing the size and shape of the pores are summarized in Table 6.3. After the centrifuge model test, the decrease of $\bar{A}$ and $\bar{D}_E$ was in coincident with the changes of the void ratio of each soil layer. The increase of $\bar{e}$ and $\bar{R}$ resulted from the pores being squashed. The increase of the average shape factor and the decrease of the average long axis length indicated that parts of the pore boundaries were compressed to be closed after the consolidation.

Figure 6.7 illustrates the pore size distribution of the silty clay of layer No. 4 and the clayey soil of layer No. 8 before and after the consolidation. The pore area of the pores with radii bigger than 1 μm occupied the majority, from 99.3 to 97.3% for the former and from 98.6 to 97.5% for the latter before and after consolidation, respectively, and the convergence was good. So the pore area of the pores with radii bigger than 1 μm can be used to represent the total pore area approximatively. After the consolidation, the total pore area of the silty clay of layer No. 4 and the clayey soil of layer No. 8 decreased from 2212 and 1388 to 1592 and 977 μm^2, respectively. The amplitudes of reduction were 28.0 and 29.6%, respectively. This

Fig. 6.7 Pore size distribution before and after consolidation

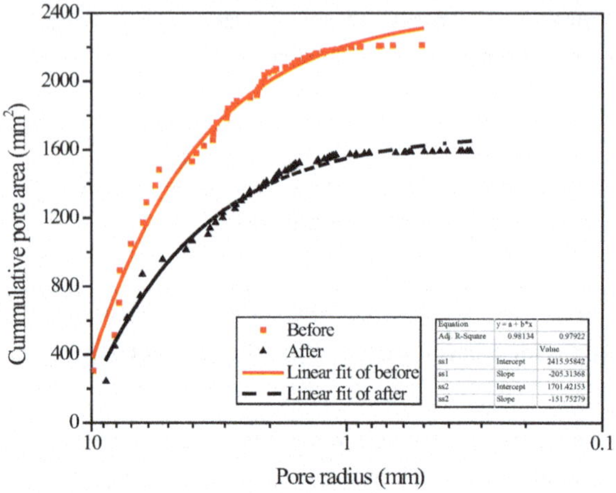

(a) Silty clay of layer No. 4

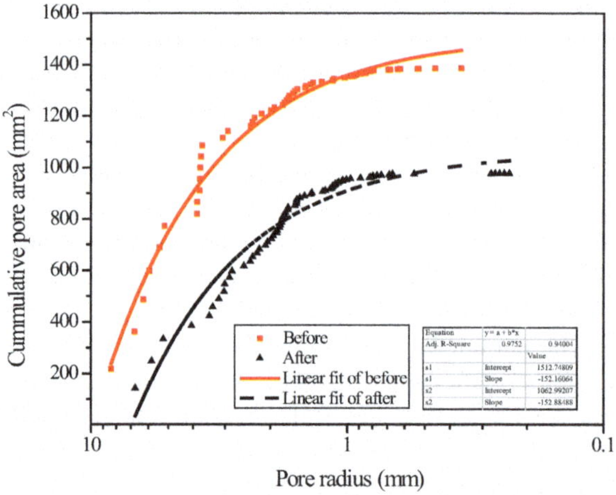

(b) Clayey soil of layer No. 8

indicated the great compressibility of the two soil layers and was in good agreement with the large macroscopic subsidence.

The silty clay of layer No. 4 and the clayey soil of layer No. 8 were compacted. The pores became smaller and parts of the pore boundaries were compressed to be closed. The shape of the pores should be more regular, which indicated that the fractal dimension of shape D should decrease. D can be calculated by

$$\log(p) = \frac{D}{2} \times \log(A) + C, \tag{6.2}$$

where p is the equivalent perimeter of a certain pore, D being the fractal dimension of shape, A being the pore area of a certain pore, and C being the constant.

Figure 6.8 shows the variations of the pore area with the perimeter of the silty clay of layer No. 4 and the clayey soil of layer No. 8 before and after consolidation. After the consolidation, D decreased from 1.448 to 1.164 for the former and from 1.341 to 1.212 for the latter. The amplitudes of reduction were 24.7 and 9.6%, respectively. This indicated that before the consolidation, the shapes of the pores of the silty clay of layer No. 4 were more complicated. However, comparing with the clayey soil of layer No. 8, the silty clay of layer No. 4 can be compacted more easily for the greater amplitudes of reduction of the fractal dimension of shape.

(2) The pore orienting characteristics

The orientation of soil microstructure influenced the soil engineering properties deeply. The orienting frequency $F_i(\alpha)$ and the orienting probability entropy H_m were used to study the pore orientation.

Figure 6.9 shows the orienting frequency of the long axis of the equivalent ellipses of the silty clay of layer No. 4 and the clayey soil of layer No. 8 before and after the consolidation. The images were centrosymmetric, and the pore anisotropy was obvious. The orienting probability entropy H_m of each image can be calculated by

$$H_m = -\sum_{i=1}^{36} F_i(\alpha) \log_{36} F_i(\alpha), \tag{6.3}$$

where $F_i(\alpha)$ is the orienting frequency of the ith range.

After the consolidation, the orienting probability entropy H_m of the silty clay of layer No. 4 and the clayey soil of layer No. 8 increased from 0.9450 to 0.9731 and from 0.9507 to 0.9569, respectively. The bigger the value of H_m, the more disordered the pores. After the consolidation, H_m of each soil layer increased and the orientation of pores became more disordered. The soils were compressed and the pores became smaller. Some longer pores were even compressed to form more pores, resulting in the dispersal of the orientation of the long axis.

Fig. 6.8 Relationship of pore area and perimeter before and after consolidation

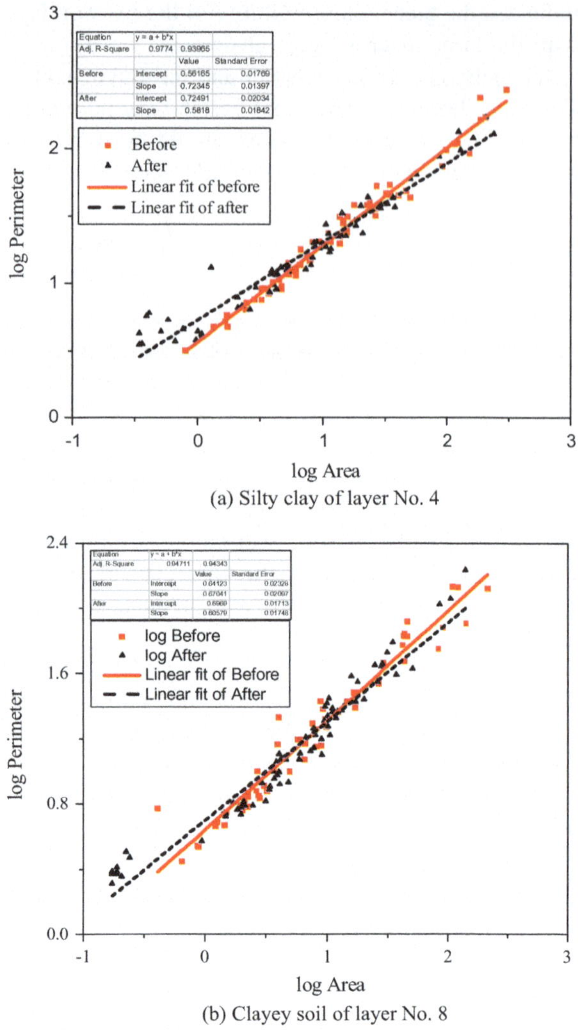

6.3 The Mercury Intrusion Porosimetry Test

6.3.1 Theory of MIP

Mercury intrusion porosimetry is based on the premise that a nonwetting liquid (one having a contact angle greater than 90°) will only intrude capillaries under pressure. MIP as a method measuring pore size distribution of porous solids was first proposed by Washburn (1921).

In order to interpret pressure values, they are usually converted into equivalent pore radii by applying the Washburn equation (Washburn 1921)

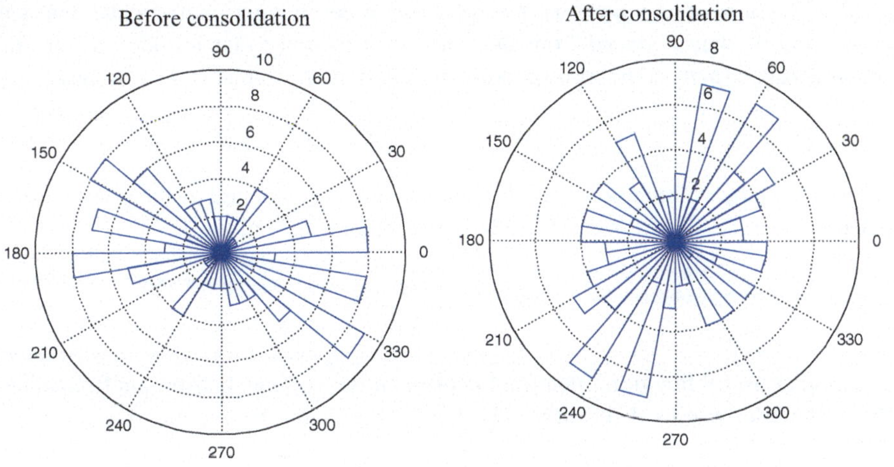

(a) Silty clay of layer No.4

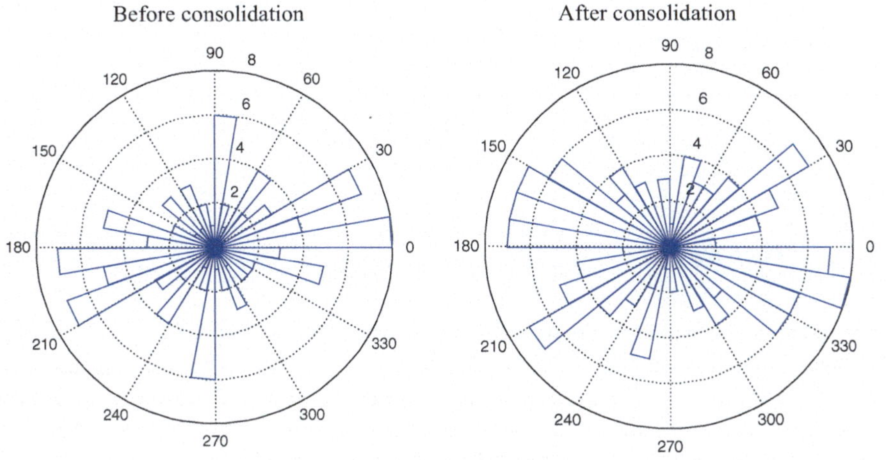

(b) Clayey soil of layer No. 8

Fig. 6.9 Orienting frequency of the long axis of the equivalent ellipses before and after consolidation

$$r = -\frac{2\gamma \cos \theta}{p}, \tag{6.4}$$

where θ is the contact angle between the mercury and a surface of the solid material tested, and γ is the surface tension of mercury. There is good agreement as to the surface tension of mercury with accepted values ranging from 0.473 to 0.485 N/m. Values for the contact angle range from 117° to 141°. These values are strongly affected by the purity of the mercury (Moro and Bohni 2002).

The fraction of the porosity occupied by pores having radii in the interval $(d, d + \Delta d)$ is then deduced from the volume of mercury that intrudes within the pressure range $(p, p + \Delta p)$ corresponding to a pore radius range $(r, r + \Delta r)$ given by

$$\mathrm{d}V = -D_V(r)\,\mathrm{d}r, \tag{6.5}$$

where D_V is the volume pore size distribution function, defined as the pore volume per unit interval of pore radii. Combining (6.4) and (6.5) gives

$$D_V(r) = \frac{p}{r}\left(\frac{\mathrm{d}V}{\mathrm{d}p}\right). \tag{6.6}$$

Another useful function often used in place of the volume distribution function is the volume log radius distribution D_V $(\log r)$, which can be expressed as

$$D_V(\log r) = \frac{\mathrm{d}V}{\mathrm{d}\log r} = rD_V(r). \tag{6.7}$$

Combining Eq. (6.4) with Eq. (6.1) yields

$$D_V(\log r) = p\left(\frac{\mathrm{d}V}{\mathrm{d}p}\right) = \frac{\mathrm{d}V}{\mathrm{d}\log p}. \tag{6.8}$$

6.3.2 *Test Result*

The mercury porosimeter typed Autoscan 60 was used in the test. The soil samples of layer No. 4, layer No. 7, layer No. 8, and layer No. 9 before and after the centrifuge model test were prepared. The mercury was intruded into the sample tube under the vacuum condition. The low-pressure analysis (Fig. 6.10a) was conducted, and then the high-pressure analysis (Fig. 6.10b) was conducted.

In the test, the surface tension of mercury was 0.485 N/m, and the contact angle of mercury was 130°. Ink-bottle effect exists in the intrusion stage. MIP involves intruding the pore space of a given material with the mercury under increasing absolute pressures. The intrusion pressure is inversely related to the equivalent pore entry diameter.

Figure 6.11 shows variations of pore size with intrusion pressure of each soil layer before and after the centrifuge model test. Porosimetry equipment can typically measure a maximum pressure range between 0.53 and 60,000 Psi. This pressure range corresponds to pore sizes from several hundreds of micrometers (low pressure) to a few nanometers (high pressure).

Figure 6.12 shows the variations of intrusion–extrusion volume of mercury with intrusion–extrusion pressure of each soil layer before and after the centrifuge model test. In the intrusion stage, the intrusion volume of mercury increases with the

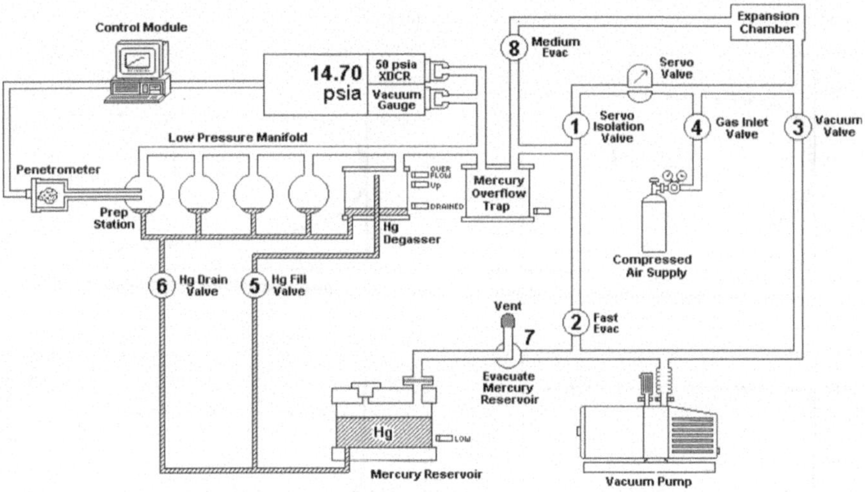

(a) Low pressure

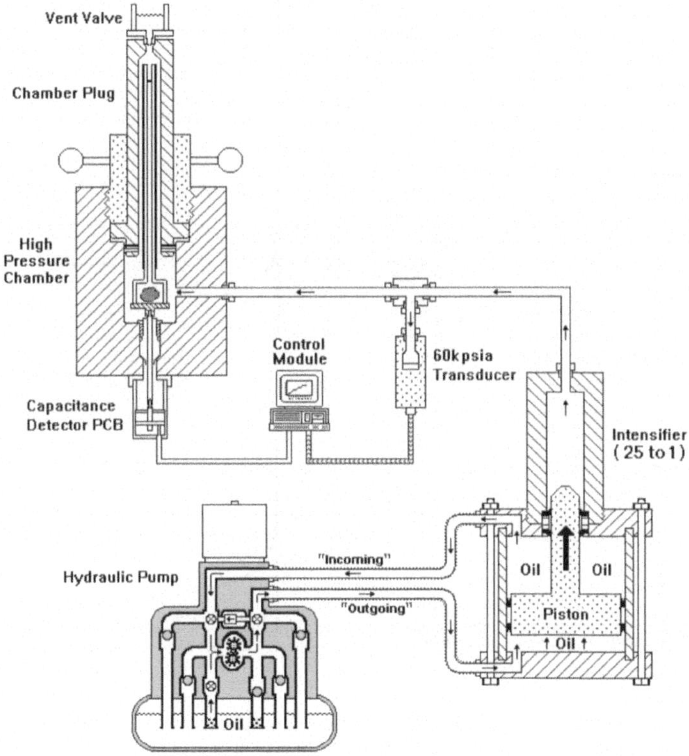

(b) High pressure

Fig. 6.10 Analysis system for low pressure and high pressure

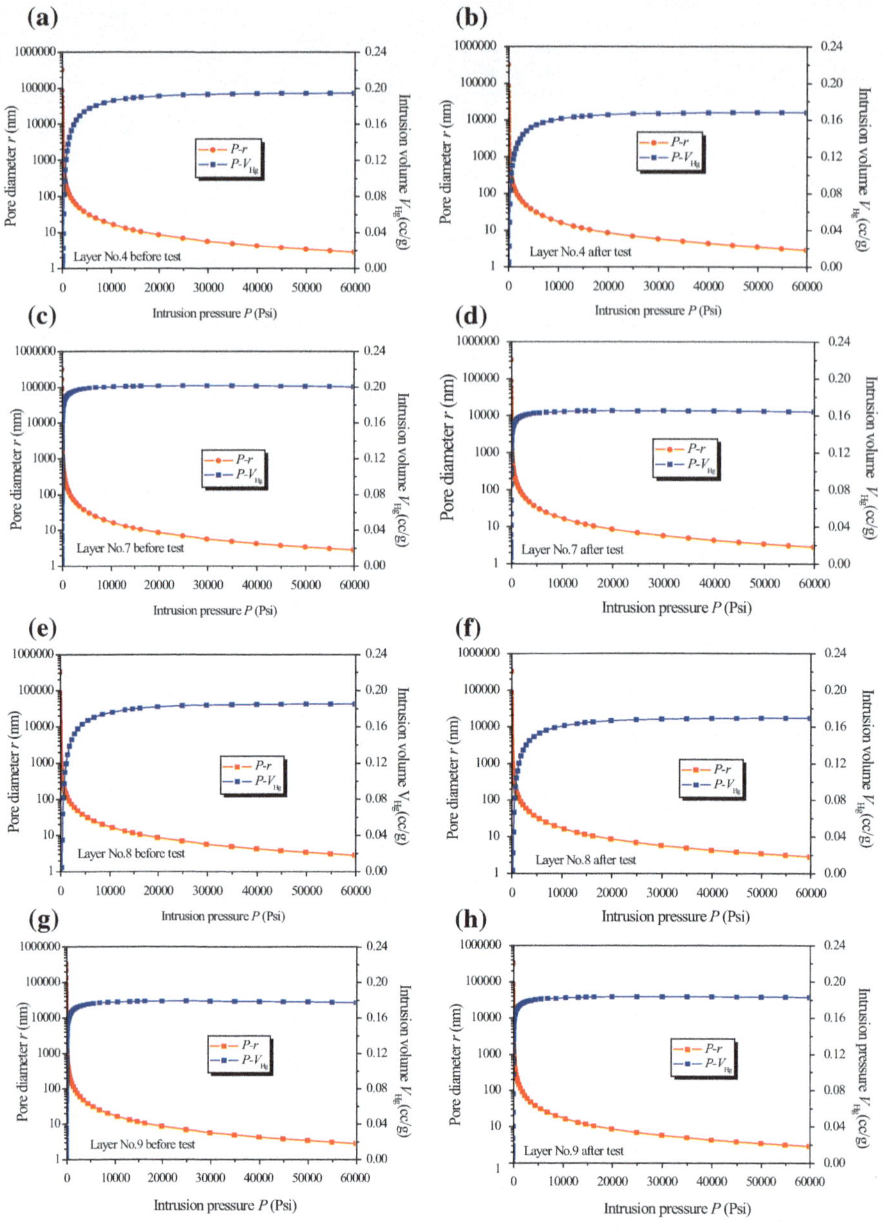

Fig. 6.11 Variations of pore size with intrusion pressure of each soil layer before and after the centrifuge model test

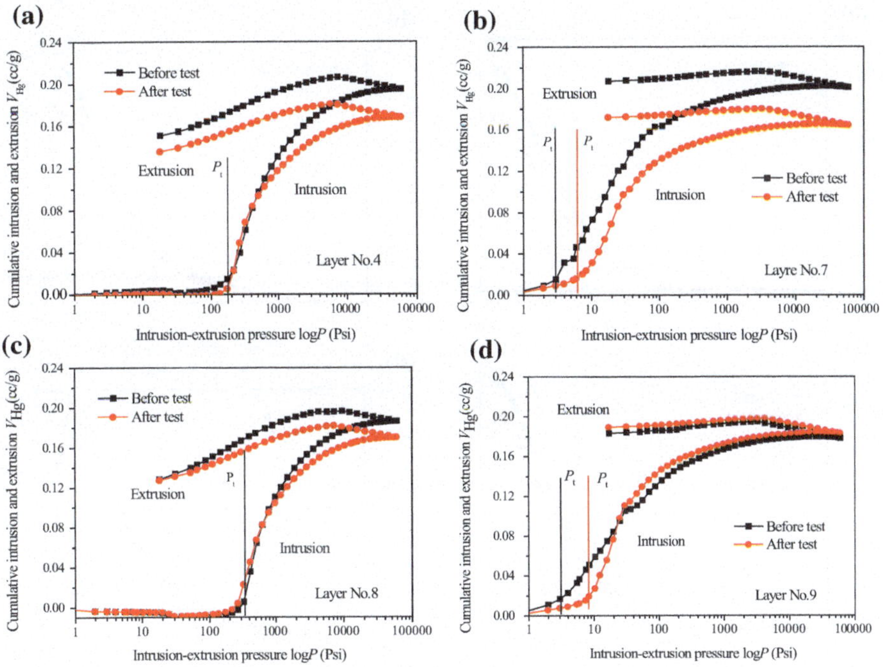

Fig. 6.12 Variations of intrusion–extrusion volume with intrusion–extrusion pressure of each soil layer before and after the centrifuge model test

intrusion pressure increasing, and the pore size becomes smaller and smaller. The intrusion volume increases slowly in the beginning. But when the intrusion pressure reaches a certain value (that is, threshold value P_t), the intrusion volume increases quickly. That means the ink-bottle effect exists in the intrusion stage. Until it reaches the peak value, it becomes smooth. The threshold value of each soil layer is different. The extrusion curve does not coincide with the intrusion curve because there is some mercury remained in the pores of soils.

Figure 6.13 shows the variations of cumulative specific surface area with intrusion pressure of each soil layer before and after the centrifuge model test. In the beginning, the cumulative specific surface area increases slowly with the intrusion pressure increasing. When the intrusion pressure reaches the threshold value, it increases stably. The cumulative specific surface area is affected by not only the pore size but also the pore number.

It is defined for soils that the pores with pore radii larger than 50 nm are macropores, pore radii between 2 and 50 nm being mesopores, and pore radii less than 2 nm being micropores. By the intrusion curve, the ratio of the volume of macropores to the volume of total pores can be obtained for each soil layer before and after the centrifuge model test, as shown in Fig. 6.14.

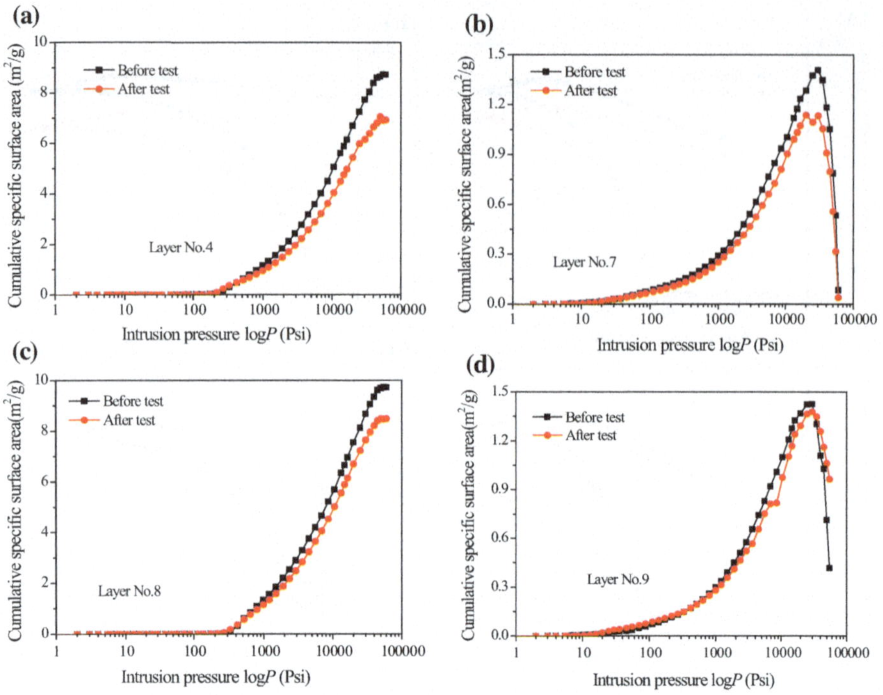

Fig. 6.13 Variations of cumulative specific surface area with intrusion pressure of each soil layer before and after the centrifuge model test

The pore volume and its distribution of soils were determined by the pressure-intrusion data, as summarized in Table 6.4. There are mainly macropores in silty clay of layer No. 4, clayey soil of layer No. 8, silty sand of layer No. 7, and layer No. 9 in Shanghai. The macropores of each soil layer larger than 50 nm after the centrifuge model test increase a little. The total porosity n_p of layer No. 4, layer No. 7, and layer No. 8 decreases after the centrifuge model test and that of layer No. 9 increases. The mean pore size r_m of layer No. 4 and layer No. 8 increases after the centrifuge model test and that of layer No. 7 and layer No. 9 decreases. The holding mercury coefficient R_f of layer No. 4 and No. 8 increases and that of layer No. 7 and No. 9 changes little. The pore specific surface area S of each soil layer increases after the centrifuge model test.

6.3.3 Fractal Characteristics of Pore Structure of Soils

The fundamental of fractal geometry was described by Mandelbrot (1983). Peitgen and Saupe (1988) presented an exposition of fractal principles; Moore and Krepfl (1991) described the application of fractal principles to the computer synthesis of

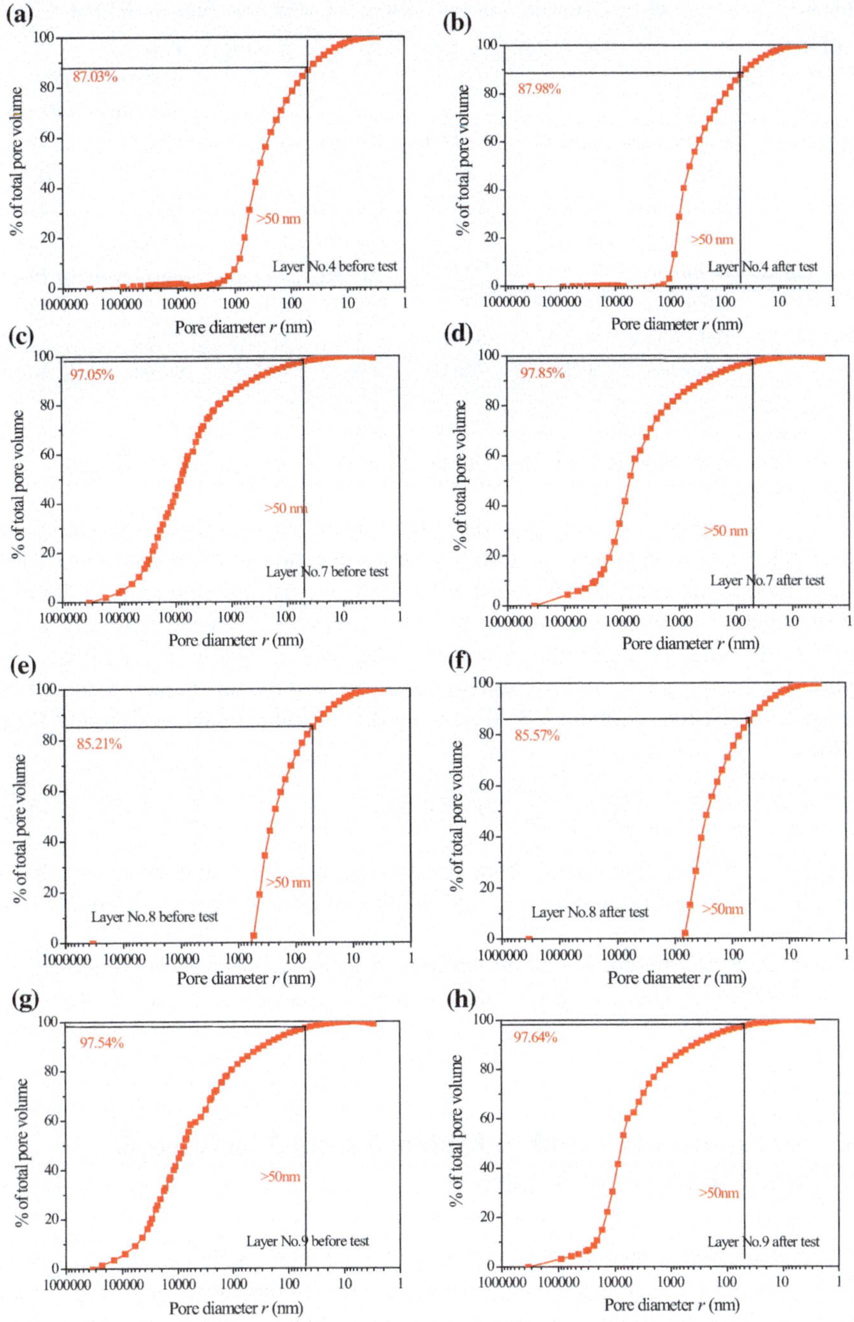

Fig. 6.14 Pore size distribution before and after the centrifuge model test

Table 6.4 Parameters of pore structures of soils before and after centrifuge model test

Soil layers		n_p (cc/g)	r_m (nm)	R_f	S (m^2/g)	Pore size distribution (%)	
						>50 nm	<50 nm
Layer No. 4	Before test	0.3003	348.4	0.7310	8.7245	87.03	22.97
	After test	0.2737	432.4	0.7522	7.0616	87.98	22.02
Layer No. 7	Before test	0.3003	8231.7	0.9563	1.4091	97.05	2.95
	After test	0.2617	7621.0	0.9536	1.1382	97.85	2.15
Layer No. 8	Before test	0.2882	245.5	0.6530	9.7245	85.21	4.79
	After test	0.2696	271.3	0.7008	8.4927	85.57	4.43
Layer No. 9	Before test	0.2701	8149.7	0.9412	1.4267	97.54	2.46
	After test	0.2807	7763.2	0.9547	1.3799	97.64	2.36

soil fabrics, and Moore and Donaldson (1995) used fractals to quantify soil microstructure.

The pore structures of many materials are of fractal characteristics (Satya Sai and Krishnaiah 2005; Mahamud et al. 2003; Thompson et al. 1987; Zhang and Li 1995; Sidney 2000; Sean et al. 2003), so are soils. If there exist line segments in the pore size distribution between the ratio of the pore volume $v(r)$ of the pore size larger than r to the total pore volume v_t and r in the double logarithm coordinate, the relationship between $v(r)$ and r in the line segment is of fractal characteristics. If the slope ratio of the line segment is b, the fractal dimension is given by Tan and Kong (2006)

$$D = 3 - b. \tag{6.9}$$

Figure 6.15 shows the fractal characteristics of pore size distribution of each soil layer after the centrifuge model test, and the fractal dimensions are summarized in Table 6.5.

There are four different fractal dimensions in silty clay of layer No. 4 and clayey soil of layer No. 8. There are three different fractal dimensions in silty sand of layer No. 7 and layer No. 9.

6.4 Mechanism of Land Subsidence from Macroscopic and Microscopic Aspect

The characteristics of land subsidence caused by the high-rise building group are as follows: silty clay of layer No. 4 experiences the maximum subsidence. The central area of the building group has larger subsidence, and the subsidence superimposition effect is obvious. It can exceed the allowance and cause land subsidence hazard; the building distance is smaller; and the subsidence superimposition effect

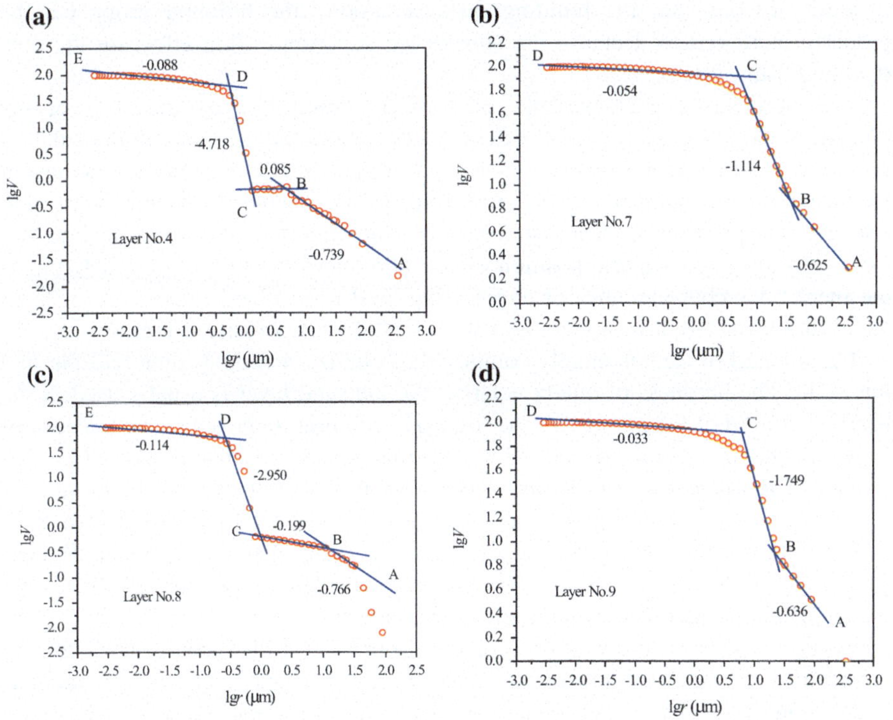

Fig. 6.15 Fractal characteristics of pore size distribution of each soil layer after the centrifuge model test

Table 6.5 Fractal dimensions of pore structures of soils after the centrifuge model test

Soil layers	Line segment	r (nm)	b	D	D_m (mean)	Specific subsidence $\bar{S}$
Layer No. 4	AB	5623–316228	−0.739	3.739	4.365	0.0092
	BC	1413–5632	0.085	2.915		
	CD	562–1413	−4.718	7.718		
	DE	3.2–562	−0.088	3.088		
Layer No. 8	AB	7943–316288	−0.766	3.766	4.007	0.0084
	BC	1002–7943	−0.199	3.199		
	CD	251–1002	−2.950	5.950		
	DE	3.2–251	−0.114	3.114		
Layer No. 7	AB	39811–316228	−0.625	3.625	3.600	0.0039
	BC	5623–39811	−1.114	4.114		
	CD	3.2–5623	−0.054	3.054		
Layer No. 9	AB	19953–316288	−0.636	3.636	3.806	0.0016
	BC	9772–19953	−1.749	4.749		
	CD	3.2–9772	−0.033	3.033		

is more obvious. So, the building distance among the building group can be properly increased to decrease the subsidence superimposition effect on the area (Cui and Tang 2010).

The engineering characteristics of soils are controlled by the state of the microstructure of soils to a great extent. Flocculation and honeycomb-flocculation are the main structure in silty clay of No. 4 and clayey soil of layer No. 8, and most of the structures are laminar, which gives birth to the overhead structure with high porosity. The particles are rearranged and consolidation is conducted by the building loads and the stress superimposition effect. The microstructure of clay determines the mechanics characteristics of engineering, and macro-land subsidence of layer No. 4 and layer No. 8 is larger. After the test, the total porosity decreases.

The mineral of sand is mainly composed of quartz, including some feldspar and mica, etc. The hardness of quartz is large. So, the sand layer is usually used as the bearing stratum of the pile tip for the high-rise building in the soft area. Because of better mechanics characteristics for engineering, the subsidence of layer No. 7 is smaller. The microstructure of silty sand is destroyed a little by the building load and the stress superimposition effect. After the test, the total porosity decreases.

The engineering characteristics of soils are controlled by the state of pore structure of soils to a great extent. The specific subsidence $\bar{S}$ of each soil layer after the centrifuge model test is summarized in Table 6.5.

Figure 6.16 shows the variations of mean pore size with specific subsidence in the centrifuge model test. For the silt sand (layer No. 7 and layer No. 9) which has larger pore size, the mean pore size decreases with the specific subsidence increasing, but for the clayey soil (layer No. 4 and layer No. 8) which has smaller pore size, the mean pore size increases with the specific subsidence increasing.

Figure 6.17 shows the variations of total pore specific surface area with specific subsidence in the centrifuge model test. For the silt sand (layer No. 7 and layer No. 9) which has larger pore size, the total specific pore surface area decreases with the specific subsidence increasing, and for the clayey soil (layer No. 4 and layer

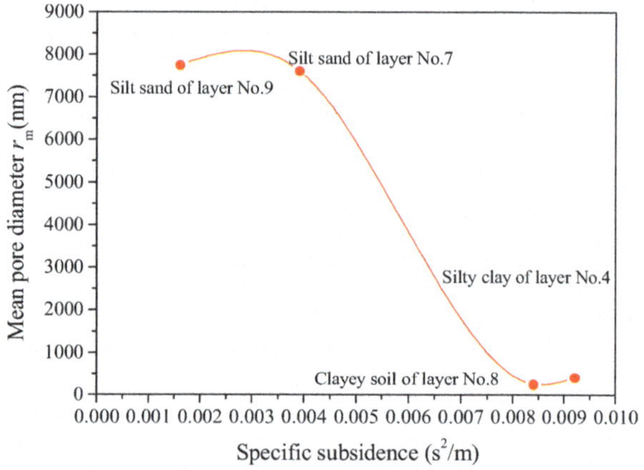

Fig. 6.16 Variations of mean pore size with specific subsidence

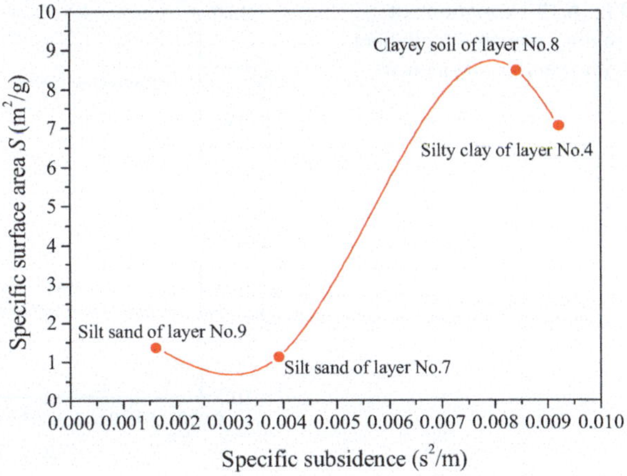

Fig. 6.17 Variations of specific surface area with specific subsidence

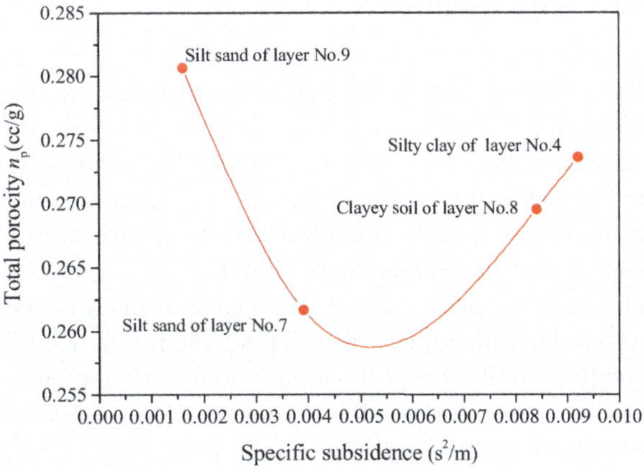

Fig. 6.18 Variations of total porosity with specific subsidence

No. 8) which has smaller pore size, the total specific pore surface area also decreases with the specific subsidence increasing.

Figure 6.18 shows the variations of total porosity with specific subsidence in the centrifuge model test, taking on the shape of saddle. For the silt sand (layer No. 7 and layer No. 9) which has larger pore size, the total porosity decreases with the specific subsidence increasing, but for the clayey soil (layer No. 4 and layer No. 8) which has smaller pore size, the total porosity increases with the specific subsidence increasing.

Figure 6.19 shows the variations of holding mercury coefficient with specific subsidence in the centrifuge model test. For the silt sand (layer No. 7 and layer No. 9) which has larger pore size, the holding mercury coefficient decreases with the specific subsidence increasing, but for the clayey soil (layer No. 4 and layer No. 8) which has smaller pore size, the holding mercury coefficient increases with the specific subsidence increasing.

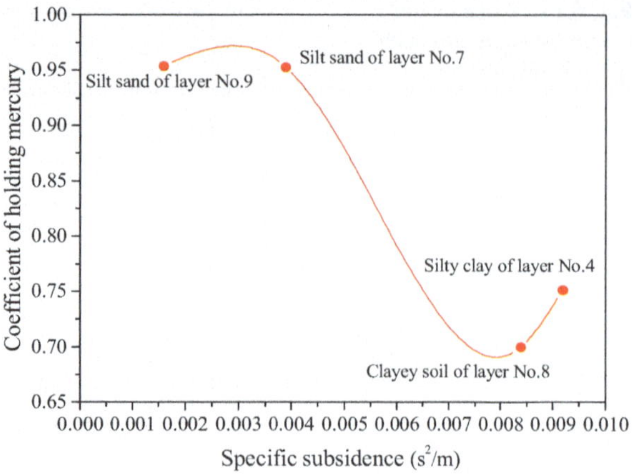

Fig. 6.19 Variations of holding mercury coefficient with specific subsidence

6.5 Discussion

This chapter studied the soil microstructure by the SEM images and the image processing effectively. It was difficult to obtain the same results as that by the laboratory test. But some results were in good agreement with the changes of the macro-parameters of soils, which can explain the process of the soil consolidation. If the void ratio calculated through the binarization of images was just the same as that of the laboratory tests, it meant that some soil particles were mistakenly identified as pores. The SEM images only illustrated the inter-aggregate pores, which determined the soil macroscopic properties such as the permeability and the compressibility, but it illustrated little intra-aggregate pores. The real filling rate n' can be estimated by adding about 15% to the calculated pore rate with the Otsu method. Another way was using a proper magnification of the images, 3000× for the fine grain or 500× for the coarse grain, and the real void ratio was estimated by thresholding the image with the threshold range [85, 90]. After the consolidation, the orienting probability entropy H_m and the disorder of the silty clay of layer No. 4 and the clayey soil of layer No. 8 increased, meaning that the soils were compressed and the pores became smaller. Some longer pores were even compressed to form more pores, which induced the dispersal of the orientation of the long axis. The increase of the average eccentricity $\bar{e}$ and the decrease of the average equivalent diameter $\bar{D}_E$ and the fractal dimension of shape D indicated that the pores were not simply squashed but parts of the boundaries of pores were compressed to be closed, which made the shape of pores more regular.

References

Cetin H, Fener M, Soylemz M, Gunaydin O (2007) Soil structure changes during compaction of a cohesive soil. Eng Geol 92:38–48

Cui ZD, Tang YQ (2010) Land subsidence and pore structure of soils caused by the high-rise building group through centrifuge model test. Eng Geol 113:44–52

Cui ZD, Tang YQ, Yan XX (2010) Centrifuge modeling of land subsidence caused by the high-rise building group in the soft soil area. Environ Earth Sci 59:1819–1826

Cui ZD, Tang YQ (2011) Microstructures of different soil layers caused by the high-rise building group in Shanghai. Environ Earth Sci 63:109–119

De Kimpe CR (1984) Effect of air drying and critical point drying on the porosity of clay soils. Can Geotech J 21:181–185

Diamond S (1970) Pore size distribution in clays. Clays Clay Miner 18:7–23

Garcia-Bengochea I, Lovell CW, Altschaeffl AG (1979) Relation between pore size distribution and permeability of silty clay. J Geotech Eng Div, ASCE 105(GT7):839–856

Hong Z, Tateishi Y, Han J (2006) Experimental study of macro-& microbehavior of natural diatomite. J Geotech Geoenviron Eng 132(5):603–610

Lapierre C, Leroueil S, Locat J (1990) Mercury intrusion and permeability of Louiseville clay. Can Geotech J 27:761–773

Lin B, Cerato AB (2012) Prediction of expansive soil swelling based on four micro-scale properties. Bull Eng Geol Environ 71:71–78

Low HE, Phoon KK, Tan TS et al (2008) Effect of soil microstructure on the compressibility of natural Singapore marine. Can Geotech J 45:161–176

Mahamud M, Loperz O, Pis JJ et al (2003) Textural characterization of coals using fractal analysis. Fuel Process Technol 81:127–142

Mandelbrot BP (1983) The fractal theory of nature. Freeman, New York

Mitchell JK (1993) Fundamentals of soil behavior, 2nd edn. Wiley, New York, pp 131–160

Moore CA, Krepfl M (1991) Using fractals to model soil fabric. Geotechnique 41:123–134

Moore CA, Donaldson M (1995) Quantifying soil microstructure using fractals. Geotechnique 45:105–116

Moro F, Bohni H (2002) Ink-bottle effect in mercury intrusion porosimetry of cement-based materials. J Colloid Interface Sci 246:135–149

Peitgen HO, Saupe D (1988) The science of fractal images. Springer, New York

Penumadu D, Dean J (2000) Compressibility effect in evaluating the pore-size distribution of kaolin clay using mercury intrusion porosimetry. Can Geotech J 37:393–405

Prapaharan S, White DM, Altschaeffl AG (1991) Fabric of field- and laboratory-compacted clay. J Geotech Eng, ASCE 117(12):1934–1940

Satya Sai PM, Krishnaiah K (2005) Development of the pore-size distribution in activated carbon produced from coconut shell char in a fluidized-bed reactor. Ind Eng Chem Res 44:51–60

Sean PR, David B, Robin SF (2003) Interpreting mercury porosimetry data for catalyst supports using semi-empirical alternatives to the Washbirn equation. Appl Catal A 238:303–318

Shi B (1996) Quantitative assessment of changes of microstructure for clayey soil in the process of compaction. Chin J Geotech Eng 18(4):57–62 (in Chinese)

Shi B, Wu Z, Chen J, Wang B (1999) Preparation of soil specimens for SEM analysis using freeze-cut-drying. Bull Eng Env 58:1–7

Sidney D (2000) An inappropriate method for the measurement of pore size distributions in cement-based materials. Cem Concr Res 30:1517–1525

Sleutel S, Cnuddeb V, Masschaelec B, Vlassenbroekc J, Dierickc M, Van Hoorebekec L, Jacobsb P, De Neve S (2008) Comparison of different nano-&micro-focus X-ray computed tomography set-up for the visualization of the soil microstructure and soil organic matter. Comput Geosci 34:931–938

Tan L (1983) The review and development of the microstructure research of the soil. Rock Soil Mech 4(1):73–86 (in Chinese)

Tan LR, Kong LW (2006) Soil science for special geotechnical engineering. Science Press, Beijing
Thompson AH, Katz AJ, Krohn CE (1987) The microgeometry and transport properties of sandstones. Adv Phys 36(5):625–694
Wang B, Shi B, Liu Z, Cai Y (2004a) Fractal study on microstructure of clayey soil by GIS. Chin J Geotech Eng 26(2):244–247 (in Chinese)
Wang LB, Frost JD, Lai JS (2004b) Three-dimensional digital representation of granular material microstructure from X-ray tomography imaging. J Comput Civil Eng 18(1):28–35
Wang B, Shi B, Cai Y, Tang C (2008) 3D visualization and porosity computation of clay soil SEM image by GIS. Rock Soil Mech 29(1):251–255 (in Chinese)
Wang C, Liu Y, Zhu Y (2011) Study on using SEM photos to obtain the pore parameters of soil samples. Saf Environ Eng 18(3):117–120 (in Chinese)
Washburn EW (1921) Note on a metrod of determining the distribution of pore sizes in a porous material. Proc Natl Acad Sci 7:115–116
Yamamuro JA, Wood FM (2004) Effect of depositional method on the undrained behavior and microstructure of sand with silt. Soil Dyn Earthq Eng 24:751–760
Zhang B, Li S (1995) Determination of surface fractal dimension for porous media by mercury porosimertry. Ind Eng Chem Res 34:1383–1386
Zhang LM, Li X (2010) Microporosity structure of coarse granular soils. J Geotech Geoenviron Eng 136(10):1425–1436

Chapter 7
Microstructures of the Soil Layer at Different Depths in the Centrifuge Modeling of Land Subsidence Caused by the Interaction of Two High-Rise Buildings

7.1 Introduction

Microlevel is one of the two points of view to investigate soil behavior of consolidation, and the MIP has been routinely used to quantitatively examine the wide range of pore sizes (Griffiths and Joshi 1989). The macrolevel behavior of soil, to a great extent, is controlled by its microstructure and mechanical properties of its particles (Cui and Jia 2013). Parameters of clay microstructure include pore size distribution (PSD), total pore volume, void ratio, etc. And the mathematical distribution of pore size in the clay can assist in interpreting the behavior of the clay (Diamond 1970). Mercury intrusion porosimetry (MIP) has been routinely used to quantitatively examine the wide range of pore sizes (Penumadu and Dean 2000). It is also used in soil study including the following: the change in permeability with compression of Louiseville clay and the relationships between pore size distribution and permeability (Lapierre et al. 1990); PSD of preconsolidated Kaolin clay using repeated MIP tests (Penumadu and Dean 2000); comparison between estimates of the PSD from the SEM and the MIP on clayey soil (Simms and Yanful 2001); variation of the inter- and the intra-aggregate pore space caused by increase in consolidation pressure of Pusan clay (Ninjgarav et al. 2007); microstructure of different soil layers in centrifuge model test and corresponding change in the PSD (Cui and Tang 2011); effects of compaction and aggregate size on the PSD of silty loam (Lipiec et al. 2012); evaluation of the PSD of cement paste on basis of pore entrapment fraction (Zeng et al. 2012); and effects of variations in moisture content on the PSD of clay (Sasanian and Newson 2013).

Initially, the adequacy of the MIP was questioned, but a number of researchers proposed the MIP to be suitable for the study of microstructure of soil. Lawrence (1978) found no significant difference in MIP tests repeated twice, which indicated that damage from the MIP was negligible. Kenney (1980) hypothesized that in spite of high pressure applied by the MIP, the force acting to cause damage was moderate, particularly in fine pores. Simms and Yanful (2002) argued the lack of

Z.-D. Cui, *Land Subsidence Induced by the Engineering-Environmental Effect*,
https://doi.org/10.1007/978-981-10-8040-1_7

variation on the PSD measured by the MIP of samples in different sizes indicated that this PSD was close to the actual PSD. Delage and Lefebvre (1984) proved that results on intra-aggregate spaces by the SEM and the MIP were of the same order of magnitude. In recent years, researchers have given their data derived from the SEM and the MIP, which are in great agreement with results of Simms and Yanful (2004).

Dehydration is an essential preparation work for MIP on clay, and the dehydration methods include air drying, oven drying, freeze drying, and critical point drying. Studies of Gillott (1970) and Mitchell and Soga (2005) revealed that air drying caused the greatest disturbance to the fabric of clay, especially for montmorillonite clays. Oven drying might be more suitable than air drying, but thermal stresses induced by it would cause substantial changes in the volume and the PSD (Delage and Lefebvre 1984; Mitchell and Soga 2005). Critical point drying is rarely used for dehydration before the MIP due to test condition restriction. Thus, it is generally accepted that freeze drying causes the minimum disturbance to clay, and it is the most appropriate choice for dehydration. It is mentionable that frozen moisture deep inside the sample may have time to change into crystalline ice and change micro-fabric of clay, which makes evaluation of sample size significant (Penumadu and Dean 2000). Shrinkage process is repeatable with oven drying; therefore, oven drying can be regarded as the most suitable dehydration method to some extent.

In this chapter, oven-drying method was used to dehydrate the clay samples due to test condition restriction. Many factors at microlevel influencing consolidation of clay were studied, and the MIP was used to examine microstructural change of Kaolin clay at different depths. The Menger fractal dimension model and the thermodynamics fractal dimension model were used to analyze the MIP data.

7.2 Test Methods

7.2.1 Centrifuge Model Test

This centrifuge model test was conducted by centrifuge in the Hong Kong University of Science and Technology. The scale of strongbox is 1250 mm 350 mm × 850 mm in length, width, and height. The test box was fronted with synthetic glass of high strength and was assembled with standard filter tubes on the bottom. Cellophanes painted with grease were stuck on the internal walls of strongbox to reduce boundary effect.

Figure 7.1 illustrates the pile foundations and bearing platform for high-rise buildings. To investigate Kaolin clay, layers in strongbox from the top downward were Kaolin clay layer and Yokohama sand layer (as bearing stratum), as shown in Fig. 7.2. The soil properties were summarized in Tables 7.1 and 7.2. The soil layer preparing procedures were as follows:

Fig. 7.1 Pile foundations and bearing platform for high-rise building model

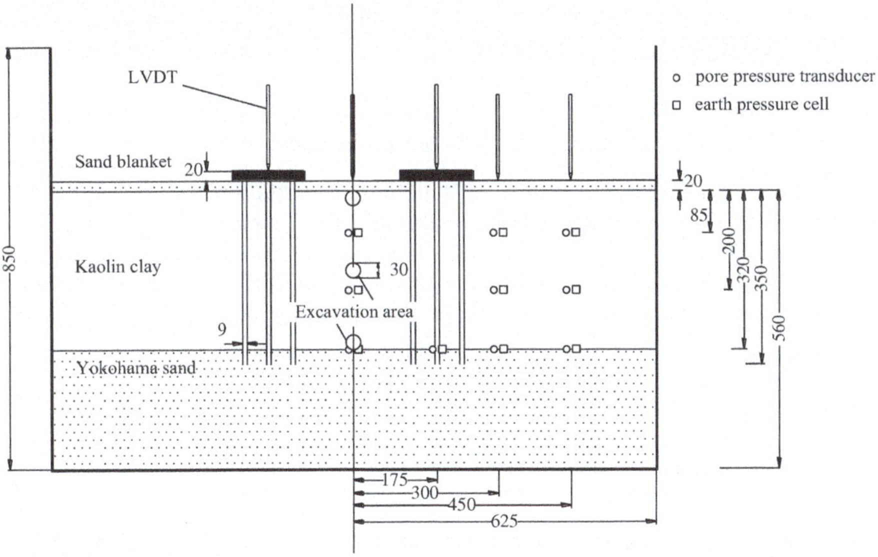

Fig. 7.2 Schematic diagram of arrangement of testing components

Table 7.1 Physical properties and component of Yokohama sand for the centrifuge model test

Soil	e_{max}	e_{min}	Relative density of soil particles (kN/m^3)	Quartz (%)	Feldspar (%)	Mica (%)
Yokohama sand	0.977	0.597	2.65	75	22	3

Table 7.2 Physical properties and component of Kaolin clay for the centrifuge model test

Soil	SiO_2 (%)	Al_2O_3 (%)	>300 mm (%)	>10 μm (%)	≤2 μm (%)	Relative density	pH
Kaolin clay	47	38	<0.02	<0.5	80 ± 3	2.6	5.0 ± 0.5
Soil	Plastic limit (%)	Liquid limit (%)	Plastics index (%)	Coefficient of permeability (m/s)	D50 (mm)		
Kaolin clay	35	60	30	$0.5–2 \times 10^{-9}$	5.0 ± 0.5		

(1) Sand rain method was used to construct the bottom Yokohama sand layer.
(2) Standard filter tubes were used to hydrate Yokohama sand layer until it was saturated. The density of layer was 1.510 kg/m^3.
(3) Kaolin clay was agitated with deionized water by vacuum mixer and turned into slurry with the water content being 110%, and then hold into the model container, 600 mm in thickness.
(4) The soil in the box consolidated in 1 g condition for 16 days, and then it was in 100 g condition to consolidate for 48 h. It was regarded as consolidated at that time.
(5) After consolidation, the Kaolin clay layer was trimmed to 320 mm in thickness. The Yokohama sand layer was 240 mm in thickness.

The model consisted of two 10-floor buildings. Two buildings with original plan dimensions 15 m × 15 m were placed on the soil. This model used pile foundations as this is the normal type of foundation for buildings. However, it is not intended to study characteristics of pile foundations, and they are used just to take building load. The centrifuge model was conducted at 100 g for 12 h. After the model test, samples were excavated as shown in Table 7.2, and the positions were circled.

7.2.2 *Procedure of the MIP Test*

Soil samples of Kaolin clay were prepared after centrifuge model test. The PSD of the samples was determined using AutoPore IV 9510 produced by Micromeritics company of the United States. Preparing and test procedure is as follows:

(1) Excavate the Kaolin clay from top, middle, and bottom of the Kaolin clay layer in the center of model of two buildings. The position of excavated Kaolin clay was shown in Fig. 7.2.
(2) Cut top, middle, and bottom samples of Kaolin clay into appropriate size of volume about 2 cm^3, and choose three unbroken samples numbered as 1, 2, and 3, respectively.
(3) Sample 1–3 were dried by oven for 24 h at 60 °C, and then determined by AutoPore IV 9510.
(4) Samples were analytically weighed and placed in the dilatometer, which were then assembled and inserted into the low-pressure station for intrusion.
(5) After low-pressure intrusion (0–21.7 psia), the samples and dilatometer were weighed again, and then inserted into the high-pressure station for intrusion.
(6) Results were analyzed after high-pressure intrusion (0.1–60,000 psia) and disposal of remaining mercury.

7.3 Test Results and Analysis

7.3.1 Intrusion and Hysteresis of Mercury in Samples

There is obviously shown in Fig. 7.3 that significant mercury intrusion occurred till the intrusion pressure gets to a certain value, which is supposed to be the intrusion pressure threshold. It is agreed by Penumadu and Dean (2000) and Cui (2008). The intrusion pressure is inversely related to equivalent pore entry diameter. In this chapter, the value of the threshold pressure is determined by the intrusion pressure corresponding to the first conspicuous incremental pore volume and summarized in Table 7.3. Other quantitative results such as total intrusion volume, porosity, median pore diameter (corresponding to the median volume), the proportion of bigger pore ($d > 100$ nm), and the largest incremental pore volume are also summarized in Table 7.3. The lower bound is determined by measurement range of AutoPore IV 9510. Figure 7.3 also shows that increasing of the depth causes decrease of the largest intrusion volume of the sample. It indicates decreasing trend of the pore size of samples, meaning more collapse of pore skeleton in deeper soil after consolidation. The decrease of total intrusion volume, proportion of bigger pore, and median pore diameter showed in Table 7.3 reveal the same trend. Lipiec et al. (2012) mentioned in their paper that the decrease of bigger pore volume always company with the increase of smaller pore volume due to compaction. It is consistent with results in this paper and explained irregular values of total porosity.

Figure 7.4 shows the hysteresis of the intrusion and the extrusion procedure. The nonlinear characteristic of the PSD curve indicates the uneven distribution of pores in Kaolin clay. The diameter of pores mostly ranges from 10 to 100 nm. During the mercury removing, some of the mercury is trapped in pores with narrow neck which leads to hysteresis. In 1989, Griffith and Joshi assumed that mercury trapped in

Fig. 7.3 Incremental of intrusion volume and its percentage with varying intrusion pressure of different samples

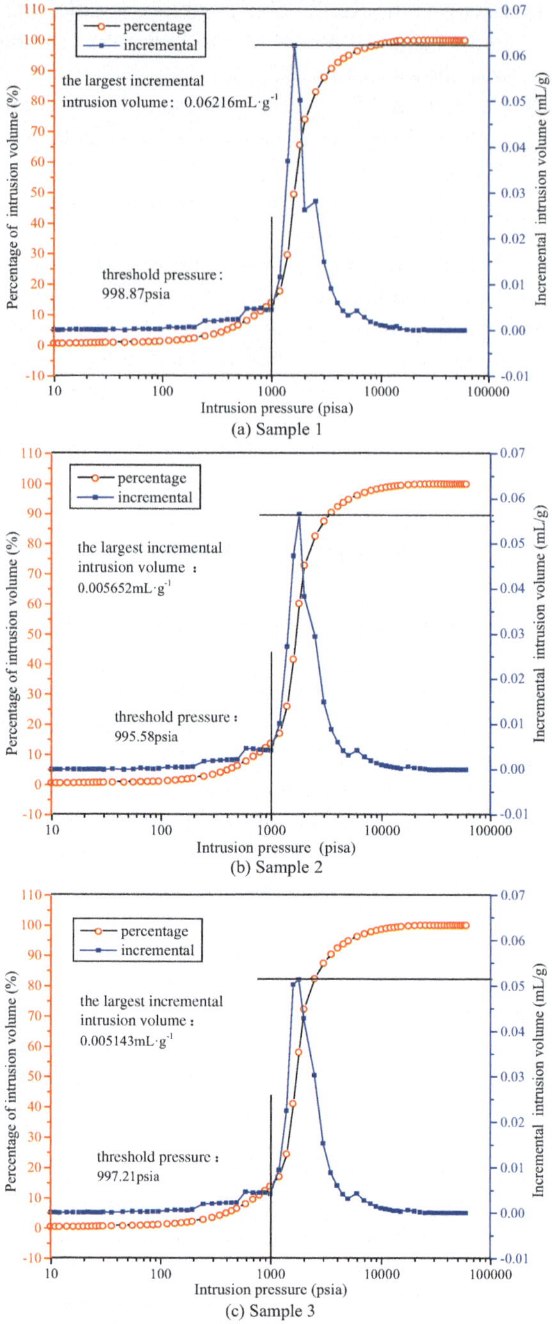

Table 7.3 Quantitative results of samples derive from MIP

Samples	Total intrusion volume (mL/g)	Largest incremental pore volume (mL/g)	Threshold pressure (psia)	Porosity (%)	Volume proportion of bigger pore (%)	Median pore diameter (nm)
1	0.3121	0.06216	998.87	39.7802	65.72	113.2
2	0.3036	0.05652	995.58	39.0140	60.24	107.3
3	0.3023	0.05143	997.21	39.0697	58.19	106.4

inter-aggregate pores and volume of free pores or intra-aggregate pores could be estimated by intrusion for two times. The principle of two-time intrusion is same as intrusion and extrusion; thus, the volume of intra-aggregate pores and inter-aggregate pores could be estimated by intrusion and extrusion procedure. The volume of trapped pores or remnant mercury stands for volume of inter-aggregate pores, and the total intrusion volume minus the volume of inter-aggregate pores stands for volume of intra-aggregate pores or free pores (Griffiths and Joshi 1989). Comparing sample 1–3, the volume of inter-aggregate pores and free pores decreases, and the proportions of volume of free pores (57.71, 58.30, and 58.32%, respectively) increase. But the variation is unconspicuous which reveals that depth of the Kaolin clay has a slight effect on particle structure.

The PSD data obtained by AutoPore IV 9510 are actually scatter. Boltzmann function fitting analysis shown in Fig. 7.5 was used to find accurate relationship between the pore diameter and the proportion of volume of pores. Fitting results are summarized in Table 7.4. It is evident that the curve fitting well with the pore diameter from 3 to 100 nm and huge numbers of pores are with diameter from 3 to 100 nm. Therefore, fitting results are practicable. Another evidence of the reliability is that the values of correlation coefficient are greater than 0.99.

7.3.2 Further Comparison of Results of Samples

The most probable aperture is corresponding to the peak value of log-differential volume of pores (Lu et al. 2011). Figure 7.6 shows the most probable aperture of each sample. Its decrease with increase of the depth indicates more collapse of pore skeleton in deeper soil after consolidation as written before. Furthermore, the most probable aperture can be a numerical standard of general pore size variation. Log-differential curves regularly show the bimodal pattern, which indicates two different mechanisms and has been used to calculate the inter-aggregate and the intra-aggregate pore volumes (Sasanian and Newson 2013). In this test, only uni-modal pattern was observed probably due to the centrifuge consolidation and the oven-drying dehydration. Compared with EPK Kaolin specimen tests carried out by Sasanian and Newson, the consolidation procedure and the oven-drying

Fig. 7.4 PSD curve of different samples in intrusion and extrusion procedure

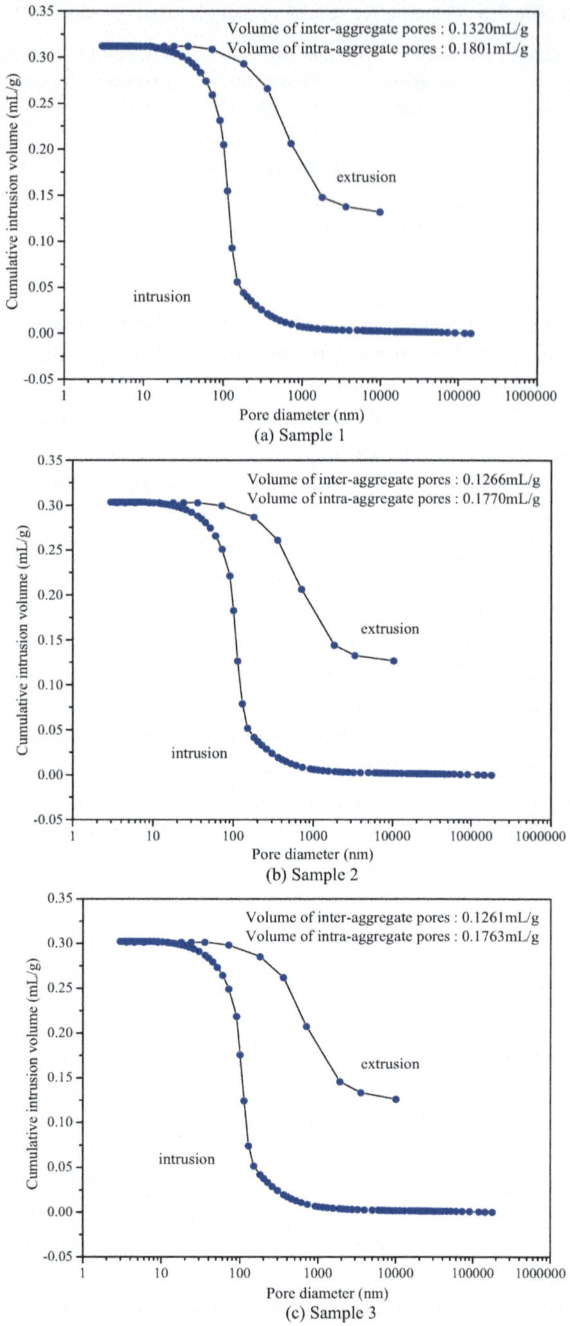

Fig. 7.5 Boltzmann function fitting for PSD of samples

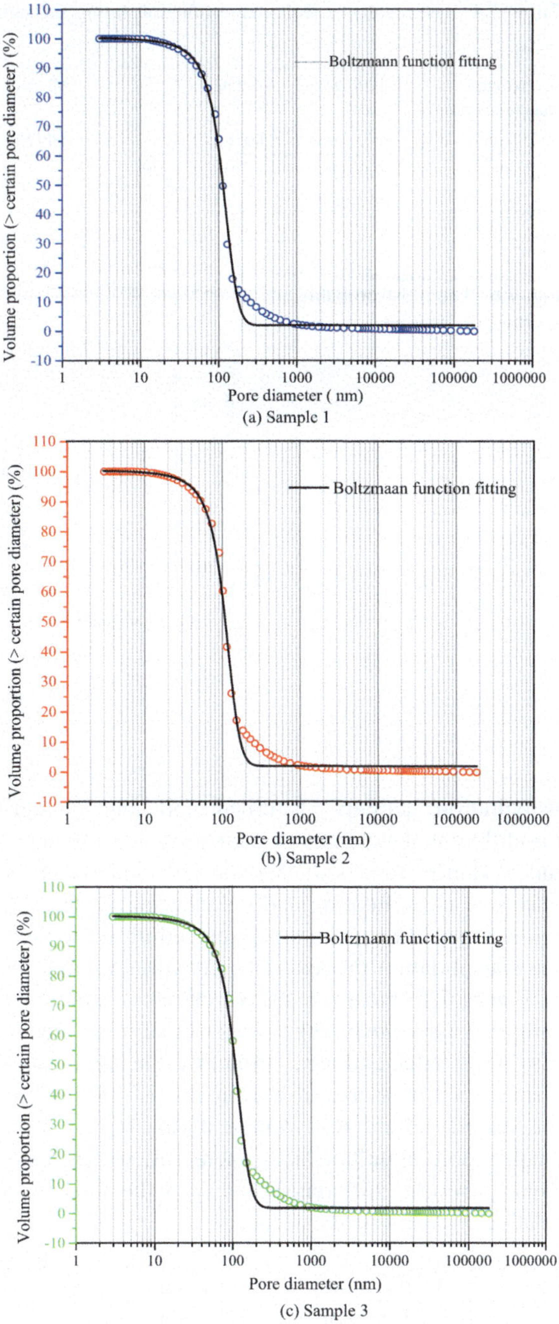

Table 7.4 The results of Boltzmann function fitting for samples

Samples		1	2	3
Boltzmann function fitting	Fitting area	All data	All data	All data
	R^2	0.99762	0.99750	0.99732
	Fitting function	$y = 1.94 + \frac{100.19}{1 + \exp^{\frac{x-111.75}{27.85}}}$	$y = 1.84 + \frac{100.18}{1 + \exp^{\frac{x-107.30}{26.44}}}$	$y = 1.84 + \frac{100.20}{1 + \exp^{\frac{x-106.28}{26.12}}}$

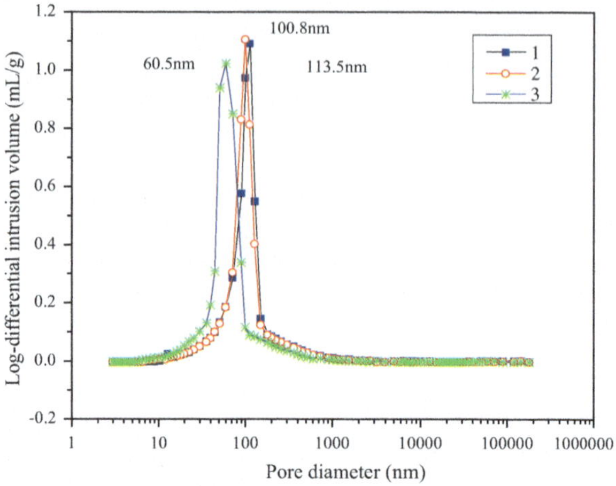

Fig. 7.6 The most probable aperture of samples

dehydration actually disturbed soil and changed the PSD of samples. But log-differential volumes in their tests and this test have the same order of magnitudes. Hence, results of this test are available. It is hard to analyze the respective effects of consolidation and dehydration on the PSD variation in this test. Further research could focus on this issue and illustrate the underlying relationship between the consolidation procedure or dehydration and the PSD variation.

Figure 7.7 illustrates the variation of cumulative pore area from sample 1 to sample 3. Obviously, the values of total pore area are not in order; furthermore, the maximum value and the minimum value are in sample 2 and sample 1, respectively. Hence, total pore area of Kaolin clay is not directly related to the buried depth. As mentioned before, the volume of the bigger pore decreases and volume of the smaller pore simultaneously increases, which leads to increase of total pore area. Total pore area differences between the three samples are not evident, meaning relatively steady of the value.

Figure 7.8 reveals the variation of the incremental intrusion volume from samples 1–3. As proposed by Simms and Yanful (2004) that the PSD of soil would change from bimodal to a unimodal distribution after compaction, the PSD of consolidated Kaolin clay only shows a unimodal pattern. It is a found law that if the variation of different similar samples is only decrease of the incremental intrusion volume peak and corresponding pore diameter to the peak is constant then both

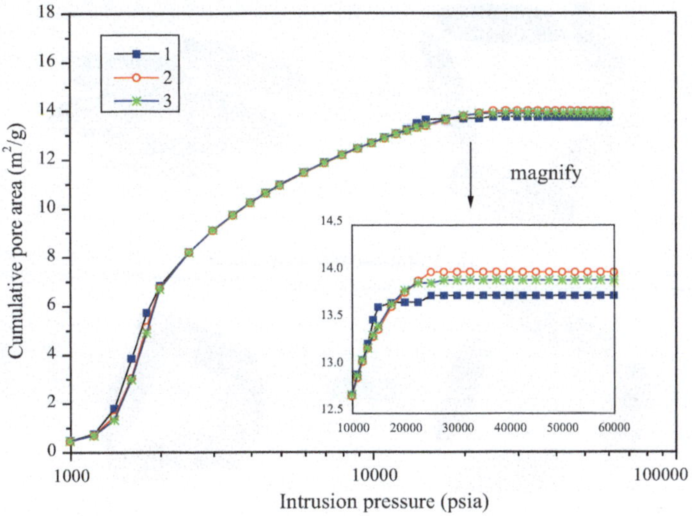

Fig. 7.7 The cumulative pore area of samples

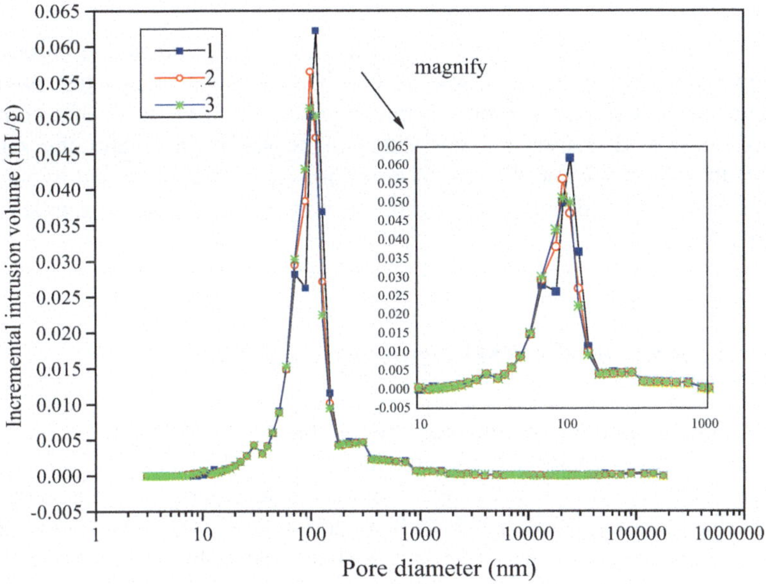

Fig. 7.8 The variation of incremental intrusion volume for samples

rotation and translation happened to soil aggregates and particles shown as Fig. 7.9a (Chen et al. 2011). In contrast to it, if the variation of different similar samples is only the decrease of the corresponding pore diameter to the peak and the

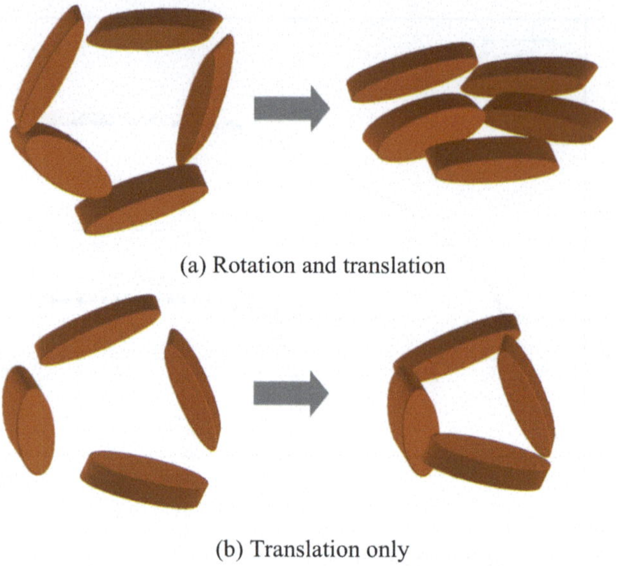

Fig. 7.9 Schematic diagram of movement of soil aggregates and particles

incremental intrusion volume peak is constant, then only translation happened to soil aggregates and particles shown as Fig. 7.9b. Results of this test obviously belong to the first situation. From sample 1 to sample 3, the peak of incremental intrusion volume decreases but the shape of curves mainly stays the same. It indicates that the shape and the size of pores are systematically varied by variation of depth in soil after consolidation; furthermore, the size of pores is smaller in soil at greater depth.

7.3.3 Analysis of the PSD Using Fractal Theory

Since Mandelbrot (1983) established the fundamentals of fractal geometry, the fractal theory was extensively used in the analysis of soil or rock microstructures. It is accepted that pore structures of soil are of fractal characteristics (Rieu and Sposito 1991; Perfect and Kay 1991, 1997; Millan and Orellana 2001; Dathe et al. 2001). Recently, Pirmoradian et al. (2005) investigated the stability of soil aggregates based on fractal theory; Wang et al. (2011) established a relationship between void ratio and fractal dimension; and Tripathi et al. (2012) were of the opinion that values of *D* reflect a strong buffering capacity of ecotone with respect to the soil aggregate formation.

Conclusions of previous researches were summarized as follows:

(1) The greater the fractal dimension of soil pores is, the more irregular the PSD is.
(2) Soil fractal characteristic reveals only in no scale area to some certain sections of data; thus, soil has multi-fractal characteristic.
(3) Fractal dimension is a quantitative coefficient of structures of soil pores and, to some extent, determines soil deformation.

In this research, Menger fractal dimension model was used in the analysis of fractal characteristics. In theory, Menger fractal dimension model was formed by making regular pores in a cube as shown in Fig. 7.10. In practical application, irregular pores are made in a cube to simulate real soil pores as follows.

If the side length of the cube is L and each side was divided into m pieces for n times, then the original cube will be composed of small cubes with the length of $l = L/m^n$, and the number of small cubes is

$$N_n = (l/L)^{-D}, \tag{7.1}$$

where $D = \lg N_{\text{end}}/\lg m$, and the volume of this porous cube is

$$V_n \sim l_n^{3-D}. \tag{7.2}$$

When $n \to \infty$, $V_n = V_{\text{end}}$, then

$$\mathrm{d}V_{\text{end}} = \mathrm{d}l \sim l^{2-D}. \tag{7.3}$$

It can be seen that the volume of pores is $V = L^3 - V_{\text{end}}$, then

$$-(\mathrm{d}V/\mathrm{d}l) \sim l^{2-D}. \tag{7.4}$$

Combining Eqs. (7.1) with (7.4), we can get the relationship between fractal dimension and intrusion pressure:

$$\lg(\mathrm{d}V/\mathrm{d}l) \sim (D - 4)\lg P. \tag{7.5}$$

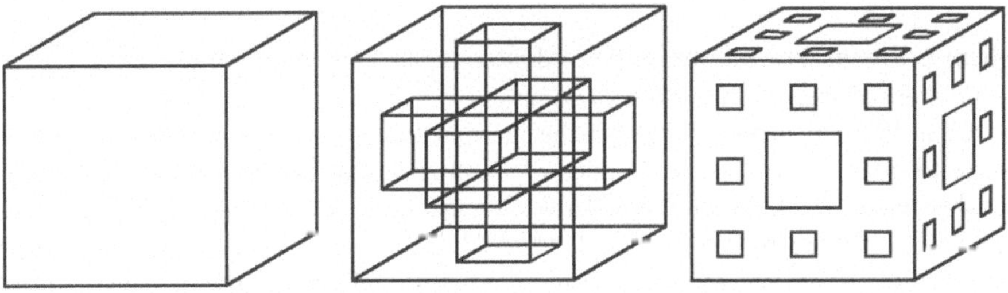

Fig. 7.10 Schematic diagram of Menger fractal dimension model

The $\mathrm{d}V/\mathrm{d}l$ and P double logarithmic curve were obtained by test data, and the pore volume fractal dimension was given according to the slope of the curve (Jiang and Lei 2010). It can be seen in Fig. 7.11 that the Menger fractal dimension model is only appropriate to sections with P from 10 to 100 MPa, and values of Menger fractal dimension are not in the reasonable range (2–3). Thus, thermal fractal dimension model was used to analyze for comparison. Comparison results are summarized in Table 7.5.

In the MIP test, the relationship between the intrusion pressure and the intrusion volume turns out to be

$$\int_0^V P\,\mathrm{d}V = -\int_0^S \gamma \cos\theta\,\mathrm{d}S, \tag{7.6}$$

where S is the area of pores.

Equation (7.6) is equivalent to

$$\int_0^V P\,\mathrm{d}V = -\int_0^S \gamma \cos\theta\,\mathrm{d}S, \tag{7.7}$$

where $\overline{P_i}$ is the mean intrusion pressure for ith scattered intrusion; ΔV_i is the intrusion volume for ith scattered intrusion; n is the total number of scattered intrusion minus 1; r_n is the pore radius for nth scattered intrusion; k is constant; and V_n is the cumulative intrusion volume for the nth scattered intrusion.

If it is given that

$$W_n = \sum_{i=1}^{n} \overline{P_i}\Delta V \tag{7.8}$$

$$Q_n = V^{1/3}/r_n. \tag{7.9}$$

Combing Eqs. (7.8) and (7.9) into Eq. (7.7) gives

$$\ln\left(W_n/r_n^2\right) = D\ \ln Q_n + C, \tag{7.10}$$

where C is constant, and the slope of double logarithmic curve of Q_n and $\left(W_n/r_n^2\right)$ is the thermal fractal dimension of samples.

It can be seen in Fig. 7.12 that the thermal fractal dimension model is appropriate to all sections, and the values of thermal fractal dimension are in the reasonable range (2–3). As shown in Table 7.5, the values of the thermal fractal fitting correlation coefficient are greater than 0.997, which means the values of thermal fractal dimension are reliable. Comparing with values of the Menger fractal dimension, the values of the thermal fractal dimension are more appropriate to

Fig. 7.11 Menger fractal dimension of samples

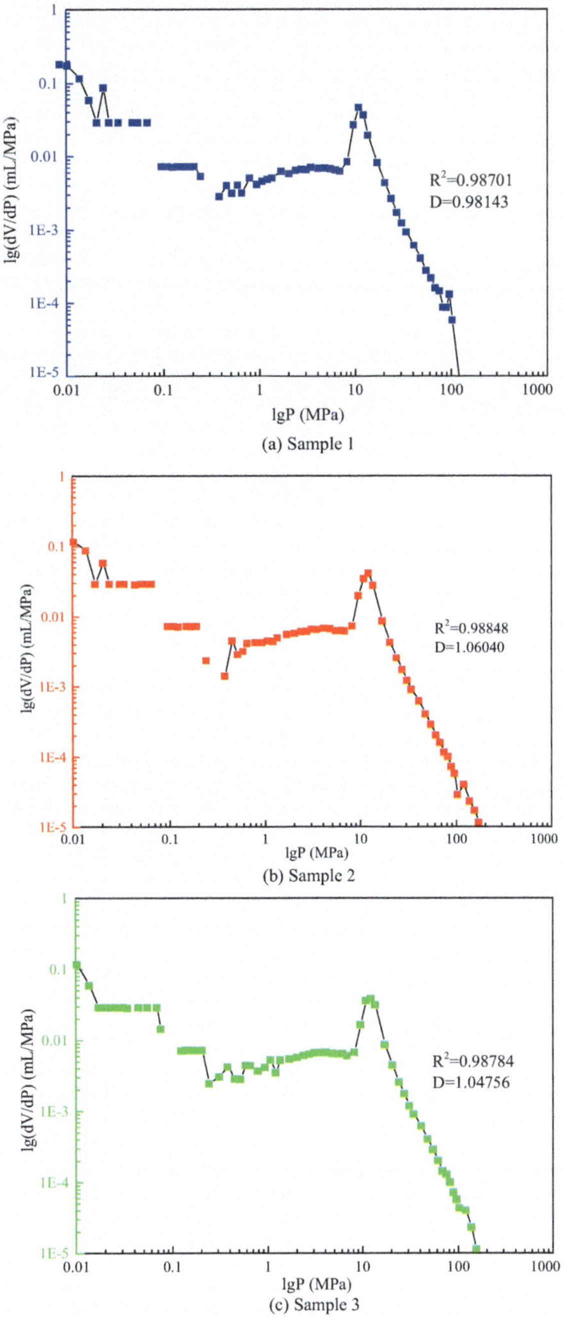

Table 7.5 Comparison of results of Menger fractal dimension and thermal fractal dimension of samples

Samples		1	2	3
Menger fractal dimension	Appropriate sections	>10 MPa	>10 MPa	>10 MPa
	R^2	0.98701	0.98848	0.98784
	D	0.98143	1.06040	1.04756
	Linear fitting function	$y = 1.69104 - 3.01857 \times x$	$y = 1.58687 - 2.9396 \times x$	$y = 1.61253 - 2.95244 \times x$
Thermal fractal dimension	Appropriate sections	All sections	All sections	All sections
	R^2	0.99708	0.99713	0.99712
	D	2.85885	2.86529	2.86751
	Linear fitting function	$y = 3.9473 + 2.85885 \times x$	$y = 3.97029 + 2.86529 \times x$	$y = 3.97747 + 2.86751 \times x$

Fig. 7.12 Thermal fractal dimension of samples

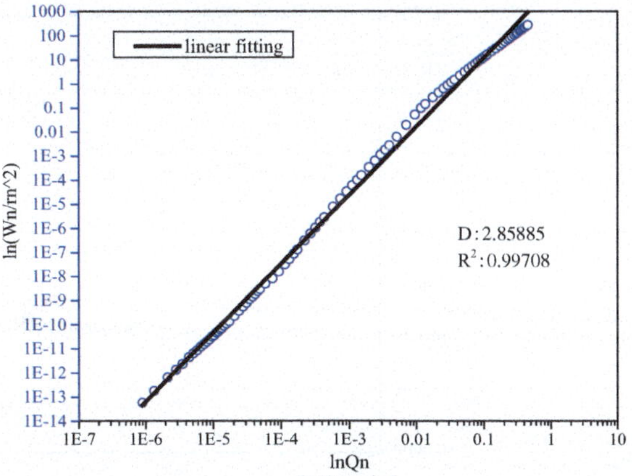

(a) Sample 1

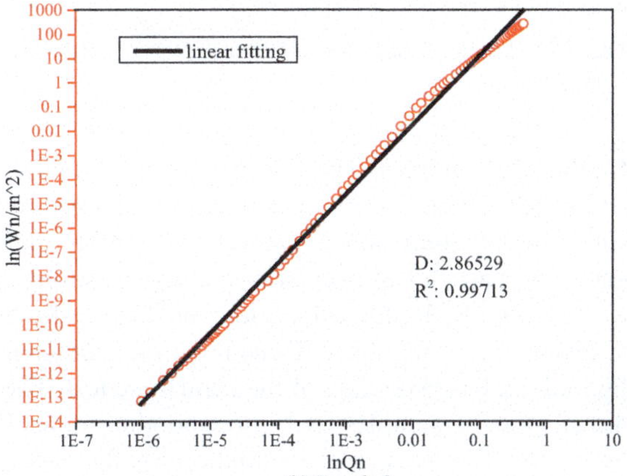

(b) Sample 2

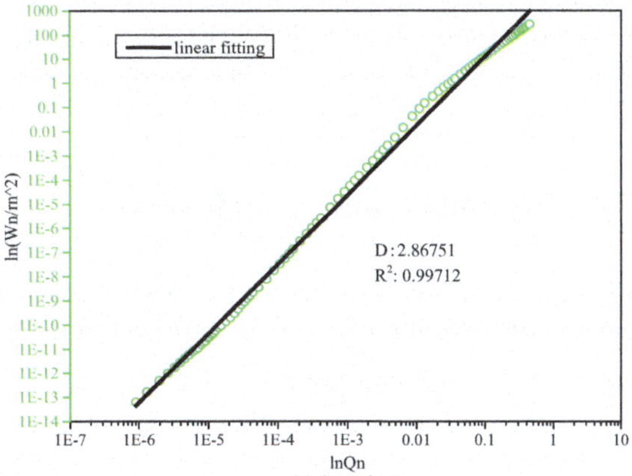

(c) Sample 3

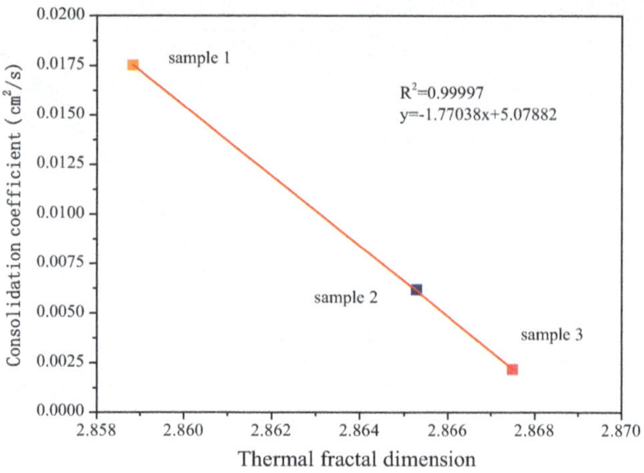

Fig. 7.13 Linear fitting for relation between consolidation coefficients and thermal fractal dimensions

Kaolin clay samples. The thermal fractal dimension increases with the increase of soil depth which reveals that pores in shallow soil are more regular than in deep soil. This phenomenon is caused by rotation and translation movement of aggregates and particles in deep soil mentioned before. In addition, the shallow soil may be disturbed by building load in centrifuge test which makes microstructure of soil to change more regularly. Consolidation coefficients in different depth of soil were figured out based on data of the centrifuge test. The relation between thermal fractal dimensions of samples and corresponding consolidation coefficients is shown in Fig. 7.13. The greater consolidation coefficients are corresponding to the greater thermal fractal dimension. Though a great linear fitting was found, it is not enough to determinately indicate the existence of the relation. So it is only a conjecture. Thus, a greater awareness of this issue must be founded on further study.

7.4 Summary and Conclusions

Mercury intrusion porosimetry test was conducted to evaluate the PSD change of consolidated Kaolin clay due to variation in depth. Conclusions are as follows:

(1) Although the threshold pressure is related to pore entry diameter, it is more closely attached to the PSD or the porosity of soil. The value of it can be estimated by variation of the incremental pore volume.
(2) Values of the total intrusion volume, peak incremental pore volume, median pore diameter, proportion of bigger pore, and most probable aperture decreased with increase in depth of soil, showing quantitative and qualitative reduction of the pore size. Values of the porosity and total pore area changed irregularly

with increase in the depth of soil, showing the corresponding increasing of smaller pores in soil.

(3) Ink-bottle effect was caused hysteresis of samples and separated the intra-aggregate pore volume from the total pore volume. Slight increase of proportions of free pore volume was observed in increasing of depth in Kaolin clay, showing the stability of clay aggregates.
(4) Rotation and translation happened to aggregates and particles of Kaolin clay at a greater depth. In addition, deeper Kaolin clay has more complicated pore structure indicated by thermal fractal dimensions, which are appropriate to Kaolin clay.

7.5 Conclusions

This chapter studied the microstructure of each soil layer under the building loads in the centrifuge model by the scanning electron microscopy and the mercury intrusion porosimetry, for qualitative analysis and quantitative analysis, respectively. The influence of the selected threshold on the calculated void ratio was studied. A method of picking up pores from the SEM images depending on the similar grayscale value was put forward. The soil microstructures of the silty clay of layer No. 4 and the clayey soil of layer No. 8 were studied by analyzing the changes of soil microstructure characteristics. The micro- and macro-mechanism of land subsidence caused by the high-rise building group was analyzed. The results are as follows:

(1) There are mainly macropores in the layer No. 4 silty clay, the layer No. 8 clayey soil, the layer No. 7 silty, sand and the layer No. 9 silty sand in Shanghai. The ink-bottle effect exists in the intrusion stage in the MIP test. There are four different fractal dimensions in layer No. 4 silty clay and layer No. 8 clayey soil. There are three different fractal dimensions in layer No. 7 silty sand and layer No. 9 silty sand in Shanghai.
(2) Flocculation and honeycomb-flocculation are the main structures in silty clay of No. 4 and clayey soil of layer No. 8, and most of the structures are laminar. The particles are rearranged, and the consolidation is conducted by the building loads and the stress superimposition effect. Layer No. 4 and layer No. 8 experience larger subsidence.
(3) The mineral of sand is mainly composed of quartz, including some feldspar and mica, etc. The sand layer is usually used as the bearing stratum of the pile tip for high-rise buildings in the soft area.
(4) For the silt sand layer (layer No. 7 and layer No. 9), the mean pore size, the total specific surface area, the total porosity, and the holding mercury coefficient decrease with the specific subsidence increasing. For the clayey soil (layer No. 4 and layer No. 8), the mean pore size, the total porosity, and the holding mercury coefficient increase with the specific subsidence increasing, but the total pore specific surface area decreases with the specific subsidence increasing.

(5) The adopted threshold determines the calculated void ratio from the SEM images. The larger the adopted threshold, the smaller the void ratio will be. The real filling rate n' can be estimated by adding about 15% to the filling rate calculated with the Otsu method.
(6) The calculated filling rate n' and the adopted threshold T_n are in good agreement with the Gaussian function relationship.
(7) When calculating the void ratio with SEM images, the appropriate amplifications of images representing the soil microstructure should be chosen: 3000× or 5000× for the fine grain soil and 500× or 1000× for the coarse grain soil. The selected T_n should be between the range [0.34, 0.36], which corresponds to the range [85, 90] of the real grayscale value.
(8) After the consolidation, the total pore area of the silty clay of layer No. 4 and the clayey soil of layer No. 8 decreased greatly, and the amplitudes of reduction are 28.0 and 29.6%, respectively, showing the great compressibility of the two soil layers, and are coincident with the large macroscopic subsidence. The pore area of the pores whose radii are bigger than 1 μm occupies at least 97% of the total pore area.
(9) Before the consolidation, the orienting probability entropy H_m of the silty clay of layer No. 4 and the clayey soil of layer No. 8 was big, meaning the large degree of the disorder of the pore structure. After the consolidation, the degree was even bigger, showing the soil consolidation process, that is, the soils were compressed and the pores become smaller. Some longer pores were even compressed to produce more pores, which induces the dispersal of the orientation of the long axis.
(10) After the consolidation, the fractal dimension of shape D of the pores in the silty clay of layer No. 4 and the clayey soil of layer No. 8 decreased. Considering the changes of the average eccentricity $\bar{e}$ and the average shape factor $\bar{F}$, it can be seen that during the soil consolidation, the pores are not simply squashed but parts of the boundaries are compressed to be closed, which makes the shape of pores more regular.

References

Chen B, Zhu R, Chang FZ (2011) Microstructural characteristics of volumetric deformation of clay under different compression stresses. Rock Soil Mech 32(1):95–99 (in Chinese)

Cui ZD (2008) Study on the land subsidence caused by the dense high-rise building group in the soft soil area. Ph.D. thesis, Tongji University, China (in Chinese)

Cui ZD, Jia YJ (2013) Analysis of electron microscope images of soil pore structure for the study of land subsidence in centrifuge model tests of high-rise building groups. Eng Geol 164: 107–116

Cui ZD, Tang YQ (2011) Microstructures of different soil layers caused by the high-rise building group in Shanghai. Environ Earth Sci 63(1):109–119

Dathe A, Eins S, Niemeyer J, Gerold G (2001) The surface fractal dimension of the soil-pore interface as measured by image analysis. Geoderma 103:203–229

Delage P, Lefebvre G (1984) Study of the structure of a sensitive Champlain clay and of its evolution during consolidation. Can Geotech J 21:21–35
Diamond S (1970) Pore size distribution in clays. Clays Clay Miner 18:7–23
Gillott JE (1970) Fabric of Leda clay investigated by optical, electron-optical, and X-ray diffraction methods. Eng Geol 4(2):133–153
Griffiths FJ, Joshi RC (1989) Change in pore size distribution due to consolidation of clays. Geotechnique 39(1):159–167
Kenney TC (1980) Discussion: frost-heaving rate predicted from pore size distribution. Can Geotech J 17:332
Lapierre C, Leroueil S, Locat J (1990) Mercury intrusion and permeability of Louisville clay. Can Geotech J 27:761–773
Lawrence GP (1978) Stability of soil pores during mercury intrusion porosimetry. J Soil Sci 29:299–304
Lipiec J, Hajnos M, Swieboda R (2012) Estimating effects of compaction on pore size distribution of soil aggregates by mercury porosimeter. Geoderma 179:20–27
Lu CF, Yuan YS, Jiang JH (2011) Effect of pore structure on gas diffusion in fly ash concrete. J China Univ Min Technol 40(4):523–529 (in Chinese)
Mandelbrot B (1983) The fractal theory of nature. Freeman, New York
Millan H, Orellana R (2001) Mass fractal dimensions of soil aggregates from different depths of a compacted Vertisol. Geoderma 101:65–76
Mitchell JK, Soga K (2005) Fundamentals of soil behavior, 3rd ed. John Wiley & Sons, USA
Ninjgarav E, Chung SG, Jang WY et al (2007) Pore size distribution of Pusan clay measured by mercury intrusion porosimetry. KSCE J Civil Eng 11(3):133–139
Penumadu D, Dean J (2000) Compressibility effect in evaluating the pore-size distribution of Kaolin clay using mercury intrusion porosimetry. Can Geotech J 37:393–405
Perfect E (1997) Fractal models for the fragmentation of rocks and soils: a review. Eng Geol 48:185–198
Perfect E, Kay BD (1991) Fractal theory applied to soil aggregation. Soil Sci Soc Am J 55:1552–1558
Pirmoradian N, Sepaskhah AR, Hajabbasi MA (2005) Application of fractal theory to quantify soil aggregate stability as influenced by tillage treatments. Biosys Eng 90(2):227–234
Rieu M, Sposito G (1991) Fractal fragmentation, soil porosity, and soil water properties: I. theory. Soil Sci Soc Am J 55:1231–1238
Sasanian S, Newson TA (2013) Use of mercury intrusion porosimetry for microstructural investigation of reconstituted clays at high water contents. Eng Geol 158:15–22
Simms PH, Yanful EK (2001) Measurement and estimation of soil-pore shrinkage in a clayey till during soil-water characteristic curve tests. Can Geotech J 38:741–754
Simms PH, Yanful EK (2002) Predicting soil-water characteristic curves of compacted plastic soils from measured pore-size distributions. Geotechnique 52(4):269–278
Simms PH, Yanful EK (2004) A discussion of the application of mercury intrusion porosimetry for the investigation of soils, including an evaluation of its use to estimate volume change in compacted clayey soils. Geotechnique 54(6):421–426
Tripathi SK, Kushwaha CP, Basu SK (2012) Application of fractal theory in assessing soil aggregates in Indian tropical ecosystems. J Forestry Res 23(3):355–364
Wang Q, Sang WF, Xu LM et al (2011) Fractal geometry of consolidation settlement of soft soil based on simulation experiment. J Jilin Univ (Earth science edition) 41(2):465–470 (in Chinese)
Zeng Q, Li KF, Teddy FC et al (2012) Analysis of pore structure, contact angle and pore entrapment of blended cement pastes from mercury porosimetry data. Cem Concr Compos 34:1053–1060

Chapter 8
Physical Model Test of Layered Soil Subsidence Considering Dual Effects of Building Load and Groundwater Withdrawal

8.1 Introduction

Land subsidence resulted in the loss of urban safe elevation, the tilt of buildings, and the damage of underground pipelines. It has been a big challenge to urban sustainable development (Shi et al. 2012; Zhang 2005). Shanghai was the first city suffering from land subsidence in China, and its economic loss from 1921 to 2000 was about 46.7 billion dollars (Zhang et al. 2003).

Before the 1960s, land subsidence in Shanghai was mainly caused by unreasonable groundwater exploitation. The groundwater was mainly extracted from the second and third confined aquifers (accounting for 86% of the total) and land subsidence developed fast (Zhang and Wei 2002). The average and maximum subsidence in urban area were 1760 and 2630 mm in 1965, respectively. Since then, comprehensive groundwater exploitation program was adopted. The groundwater was mainly extracted from the fourth and fifth confined aquifers (accounting for 85% of the total) and recharged to the second and third confined aquifers. The subsidence rate slowed down. However, from the 1990s, the subsidence rate accelerated again, and the average annual subsidence was more than four times that of the previous period.

How to explain this phenomenon? Some researchers analyzed land subsidence and net groundwater exploitation quantity in Shanghai and concluded that groundwater exploitation from the deep confined aquifer, especially the increase of groundwater exploitation during 1990s, was the main cause (Zhang 2005; Shen and Xu 2011; Chai et al. 2004; Xue et al. 2005; Wu et al. 2010; Shi et al. 2008). Some researchers attributed the subsidence acceleration to large-scale engineering construction (Xu et al. 2011; Yan et al. 2002). Some researchers considered that compression of soil at certain buried depth above was caused by engineering construction, especially by high-rise buildings, and the below by the groundwater exploitation.

Z.-D. Cui, *Land Subsidence Induced by the Engineering-Environmental Effect*,
https://doi.org/10.1007/978-981-10-8040-1_8

It is difficult to measure land subsidence in site, and the measured value is influenced by many factors. A simple way is to conduct the physical model test, because the building load and the boundary condition can be controlled easily. Centrifuge model test of land subsidence caused by the groundwater exploitation for one soil layer was conducted by Cheng et al. (2011) and Sun et al. (2008). Land subsidence caused by high-rise building group with the invariable groundwater table was studied by Tang et al. (2008a, b). Cui et al. (2010) and Liu (2010) measured the subsidence of the whole soil layer caused by the building load using the PIV, yet the deformation of each soil layer was not studied.

In this chapter, the deformation of each soil layer caused by high-rise buildings and groundwater exploitation simultaneously was studied. The physical model test based on the typical geological background in Shanghai was conducted, and PIV was adopted to measure the displacement of each point in soil layers.

8.2 Physical Model Test

8.2.1 Soil Property and Pile

The model included clay and sand. The clay was the mucky soil of the No. 4 soil layer in Shanghai, and the sand was manually sieved fine sand. Their parameters are summarized in Table 8.1. The soft soil of No. 4 layer is widely distributed in Shanghai and the thickness is large, so the pile foundation is adopted for high-rise buildings. This model test adopted pile foundation, and the sand layer was chose as the bearing stratum. It was not intended to study the characteristics of pile foundations, and they were used in the model only to take the building load.

In order to simplify the model, five aluminum bars were used to make the model piles of each building, 8 mm in diameter and 450 mm in length. The aluminum plates were used to make the model pile caps of two buildings, 200 mm in length, 200 mm in width, and 10 mm in thickness.

Table 8.1 Physical parameters of soils

Soil layers	Water content w (%)	Void ratio e	Weight density γ (kN m^{-3})	Modulus of compressibility $E_{s0.1-0.2}$ (MPa)	Cohesion c (kPa)	Internal friction angle φ (°)
Clay	40.7	1.10	18.4	3.6	8.8	4.4
Sand	17.3	0.64	19.3	16.4	0.19	39.1

8.2.2 Model Test Design

The model consisted of two high-rise buildings and five soil layers using a scale of 1–100. Figure 8.1 illustrates the soil layers, from top to bottom, which are the hard layer, the first aquitard, the first aquifer, the second aquitard, and the second aquifer. The total thickness of soil layers was 1225 mm. The top hard soil layer, 25 mm in height, was conducted to prevent water evaporation from soils, and the compression was ignored. The original plane dimension of each building is 20 m × 20 m, with 35 floors. The load of each building in the model was determined to be 25 kg, and the plane dimension is 200 mm × 200 mm. Before testing, the model box was dealt with waterproofing work. Drainage pipes were installed around the inside wall of model box for water draining when loading. Smooth plastic membrane was smeared around the inside wall of model box to minimize the side friction. The initial groundwater table was just at the soil surface, that is, the whole soil model was saturated. Earth pressure cells (EPC) and pore pressure transducers (PPT) were installed in each soil layer, and linear variable differential transformers (LVDT) were installed on the soil surface, as shown in Fig. 8.2.

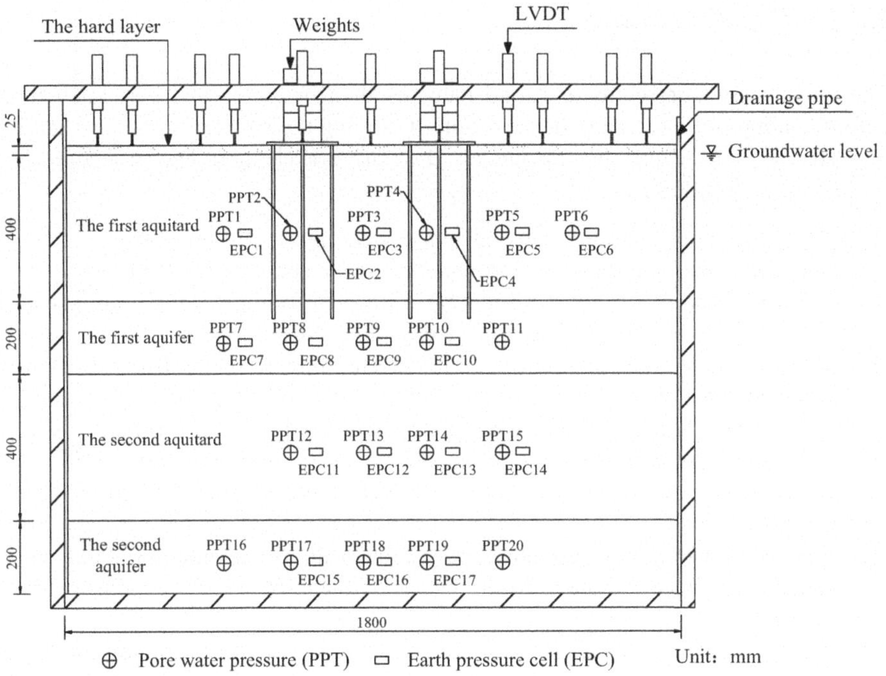

Fig. 8.1 Section view of the physical model

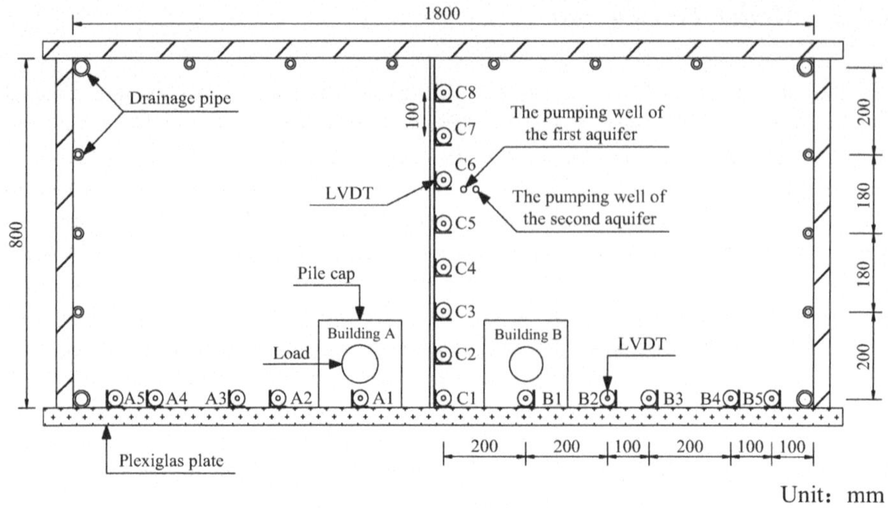

Fig. 8.2 Top view of the physical model

8.2.3 Test Procedures

The transducers were calibrated, and then installed in each soil layer. Each time before loading or dewatering, the criterion of judging the soil deformation reaching to a steady state is that the transducers data did not change. The procedures of testing were as follows:

(1) Filled the model box and installed transducers, piles, and pile caps.
(2) After the consolidation of soil completed under self-weight, Building A was loaded in three steps. First, loaded 50 N and consolidated for 1 day; second, loaded 100 N and consolidated for 1 day; and thirdly, loaded 100 N.
(3) The model was consolidated for 2d under Building A. Then, Building B was loaded in three steps. First, loaded 50 N and consolidated for 1 day; second, loaded 100 N and consolidated for 1 day; and third, loaded 100 N.
(4) After the model consolidation completed under the building load, 2500 ml of groundwater was extracted from the first aquifer. The extraction lasted 150 min.
(5) After the model consolidation completed under the first extraction, 2500 ml of groundwater was extracted from the second aquifer. The extraction lasted 150 min.
(6) After the model consolidation completed under the second extraction, 2500 ml of water was recharged to the first aquifer.
(7) After the model consolidation completed under the first recharge, 2500 ml of water was recharged to the second aquifer.

8.3 Analysis of the Test Results

8.3.1 *Subsidence at the Period of the Building Load*

8.3.1.1 Subsidence of Buildings and the Surrounding Points

Figure 8.3 illustrates the subsidence of buildings under the building load, and the period lasts 240 h. The subsidence of Buildings A and B is 5.8 and 3.9 mm, respectively. The weights of Buildings A and B are the same. The subsidence of Building A is larger than that of Building B, for Building A is constructed first. Denote the positions where Buildings A and B are built as Points A1 and B1. If the subsidence of Points A1 and B1 accumulated from the beginning of Building A is constructed, the subsidence of Point B1 is 5.2 mm, close to that of Point A1. When

Fig. 8.3 Variations of the subsidence of two buildings with time

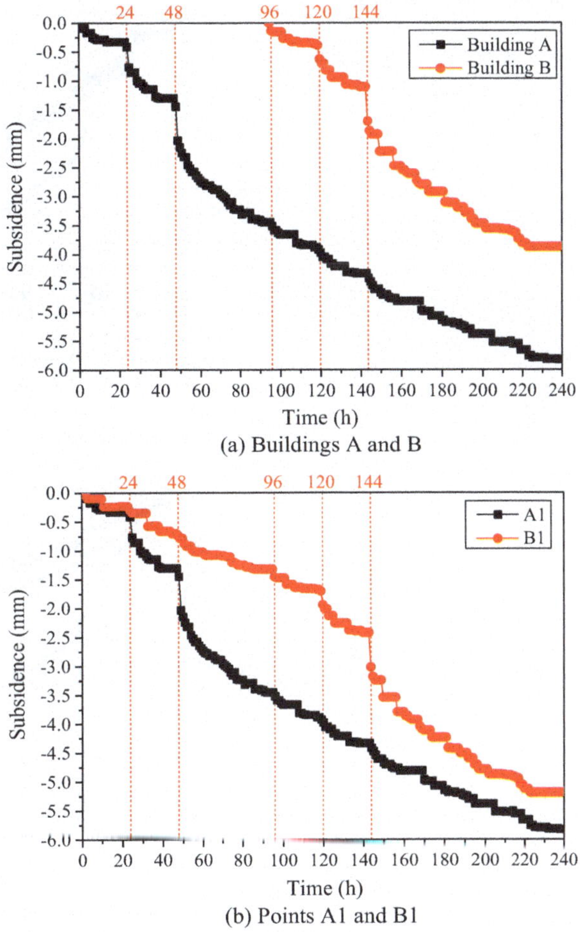

Building B is loaded, the subsidence rate of Building A increases. The earlier constructed building reduces the subsidence of the later constructed building to some extent. Land subsidence caused by buildings interacts with each other. The subsidence of Building A reaches to a steady state quickly in the first and second loading steps. During the third loading step, the subsidence keeps increasing, and the soil deformation shows the nonlinear characteristics. The subsidence rate increases abruptly when the load reaches a certain value.

Figure 8.4 illustrates the comparison of subsidence of building positions with surrounding measuring points. Points A2, B2, C1, C2, and C3 are 0.5 time the base width away from buildings. Point C1 is close to the plexiglass plate, which is laterally restrained. The subsidence of Point C1 is 4.6 mm, less than that of Points C2 and C3, which are 5.0 and 5.5 mm, respectively. The subsidence of Points C2 and C3 is close to that of Points A1 and B1 and larger than that of Points A2 and

Fig. 8.4 Subsidence of building positions and surrounding measuring points

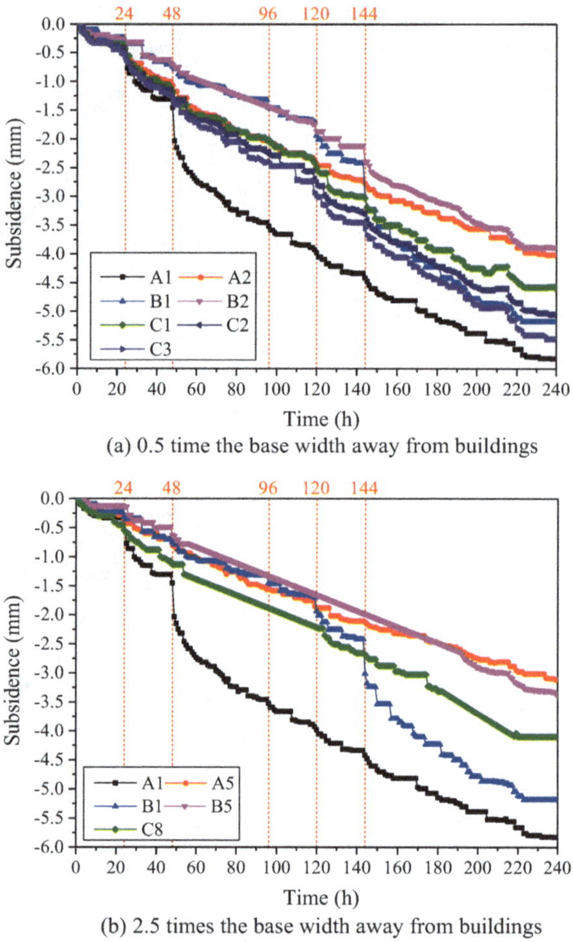

B2. This indicates that in practical engineering construction, it is necessary to pay attention to the subsidence of the central area. There are two building models in this test. Point C3 experiences larger subsidence than that of Point B1, which shows the effect of stress superimposition. Points A5, B5, and C8 are 2.5 times the base width away from buildings. Point C8 is located at the symmetry axis, and its subsidence is 4.1 mm, larger than that of Points A5 and B5, which are 3.1 and 3.4 mm, respectively. The subsidence of Points A5 and B5 is about 50% of that of the building, while that of Point C8 located at the symmetry axis is about 70%. This indicates that the subsidence of the position with a certain distance away from the buildings should also be concerned in the practical engineering construction.

8.3.1.2 Distribution of Land Subsidence

When the damage caused by land subsidence is evaluated quantitatively, subsidence value should certainly be considered. What is more, the differential subsidence may cause more economic loss. The differential subsidence may cause great additional stress to the building and the pavement structure. It may tilt high-rise buildings. So, it is necessary to study the subsidence distribution around buildings. Figure 8.5 illustrates seven typical paths which are chosen to study the subsidence distribution. Path M is located at the symmetry axis of the model. Path L is located at the top surface of the model and close to the plexiglas plate. Paths A1 and B1 are

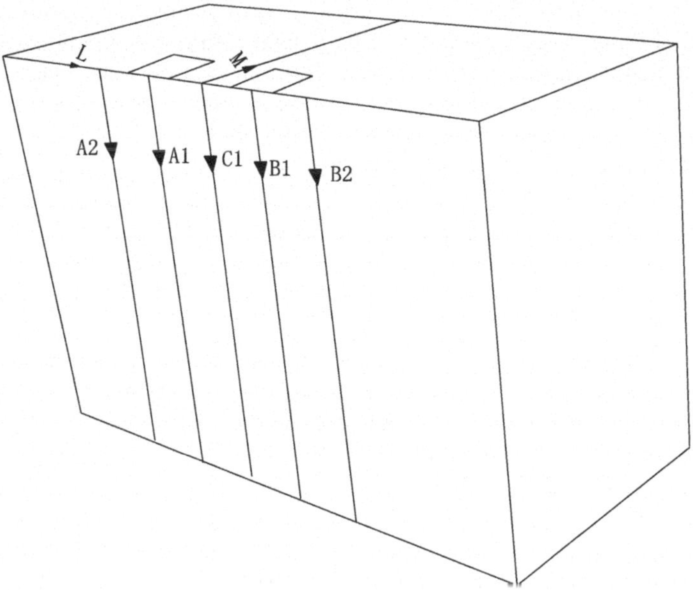

Fig. 8.5 The position of typical paths

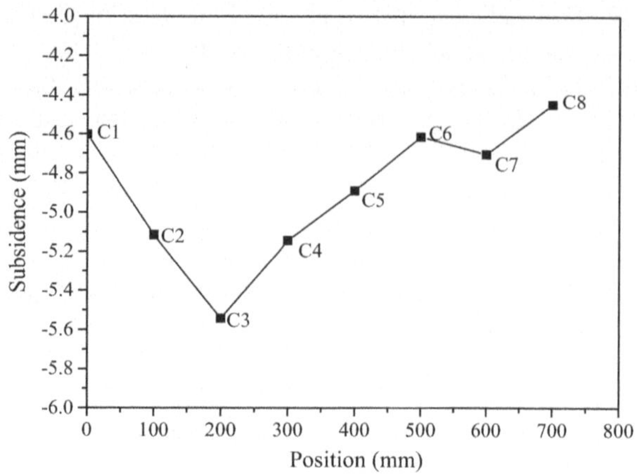

Fig. 8.6 Subsidence of points along Path M at the building load period

vertically downward the building center. Paths A2, B2, and C1 are 100 mm away from the building center.

The subsidence of 11 points (Points A1–A5, Point C1, and Points B1–B5) at Path L is measured by LVDTs. By mesh refinement, the subsidence of more points at Path L can be obtained by the PIV. However, the PIV cannot be used for Path M, and the LVDT is used to measure the subsidence of points (Points C1–C8) at Path M, while the PIV can measure the subsidence of the points at other paths. When the soil consolidation completes under the building load, the subsidence of Points C1–C8 is 4.6, 5.1, 5.5, 5.1, 4.9, 4.6, 4.7, and 4.5 mm, respectively, as shown in Fig. 8.6. Taking no consideration of the lateral restraint, for Points C3–C8, the average differential subsidence is 0.22 mm/100 mm, and the maximum differential subsidence is 0.4 mm/100 mm, located between Point C3 and Point C4.

By analyzing photos of specific section of the physical model taken at different time, the soil displacement of the whole area in the photos can be obtained by the PIV. The procedure of the PIV includes determining the coordinates of reference points, the meshing, and the displacement calculation. The grid size can be determined according to the demand for precision and data volume (White et al. 2003). Each grid in this paper is 640 pixels, which represents the size of 27 mm × 27 mm in physical space.

To validate the measured results by the PIV, the subsidence of Points A1, B1, and C1 measured by PIV is compared with that measured by the LVDT, as shown in Fig. 8.7. The subsidence of Point A1 at the end of the building load period, the first dewatering period, the second dewatering period, and the recharging period measured by the PIV is 5.6, 8.5, 11.9, and 12.8 mm, respectively, while that by the LVDT is 5.8, 8.9, 12.6, and 13.5 mm, respectively. The relative errors are 3.8, 4.6, 6.0, and 5.1%, respectively. The subsidence of Point B1 at the end of four periods measured by the PIV is 5.2, 8.7, 12.1, and 12.8 mm, respectively, while that by the LVDT is 5.3, 9.0, 12.7, and 13.4 mm, respectively. The relative errors are 1.1, 3.0, 4.5, and 4.1%, respectively. The subsidence of Point C1 at the end of the four

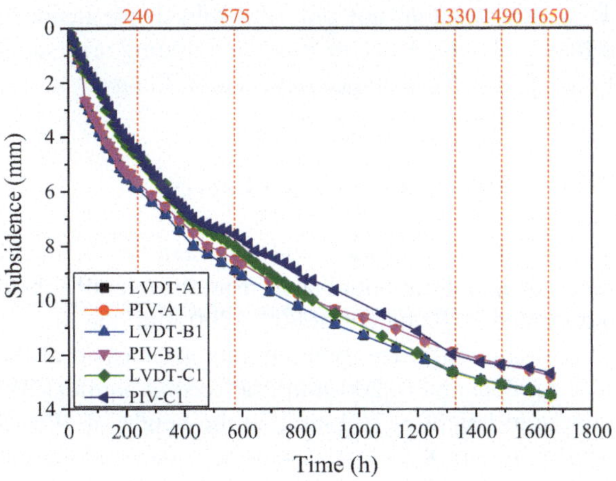

Fig. 8.7 Comparison of subsidence measured by PIV and LVDT for typical points

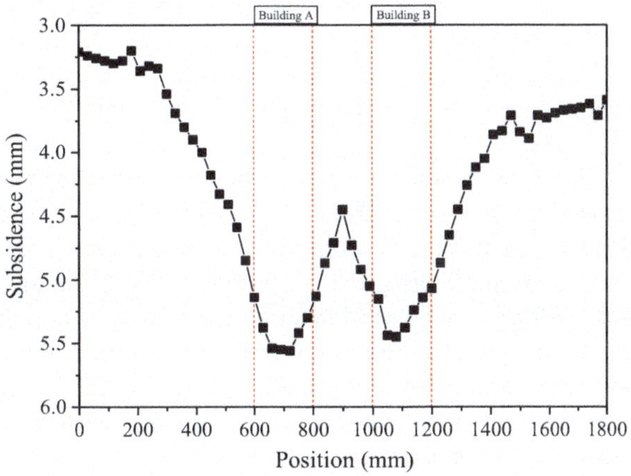

Fig. 8.8 Subsidence of points along Path L at the building load period

periods measured by PIV is 4.5, 7.6, 12.0, and 12.7 mm, respectively, while that by LVDT is 4.6, 8.0, 12.6, and 13.5 mm, respectively. The relative errors are 3.3, 5.3, 5.0, and 5.9%, respectively. The maximum relative error is about 5%. It is appropriate to study the subsidence distribution by the subsidence measured by the PIV.

Figure 8.8 illustrates the variations of the subsidence along Path L at the end of the building load period. The differential subsidence at the left of Building A and at the right of Building B is 1.0 mm/100 mm and 0.7 mm/100 mm, respectively, while that at the right of Building A and at the left of Building B is 0.5 mm/100 mm and 0.3 mm/100 mm, respectively. The differential subsidence of the area outside the building area is larger than that of the area inside the building area. The maximum differential subsidence along Path L is larger than that of Path M.

This indicates that not only the subsidence inside the building area but also the differential subsidence outside the building area should be taken into consideration to determine the distance between buildings.

8.3.1.3 Deformation of Soil Layers

The compression of each soil layer can be obtained by subtracting the subsidence recorded in an individual layer from the overall subsidence. The two aquitard layers are located at the height of 800–1200 mm (the first aquitard) and 200–600 mm (the second aquitard). The two aquifers are located at the height of 600–800 mm (the first aquifer) and 0–200 mm (the second aquifer). When analyzing the photos, the minimum height is 100 mm, so the height of 100–200 mm represents the second aquifer. Figure 8.9 illustrates the subsidence along Paths A2, A1, C1, B1, and B2 at the building load period. The compression of the first aquitard along Paths A2, A1, C1, B1, and B2 is 1.3, 1.9, 1.4, 1.7, and 0.9 mm, accounting for 34.1, 33.0, 30.8, 33.0, and 24.0% of the total subsidence, respectively. The compression of the second aquitard along the five paths is 2.3, 3.2, 2.7, 3.0, and 2.7 mm, accounting for 59.0, 56.0, 60.4, 58.0, and 69.0% of the total subsidence, respectively. The compression of the two aquitards accounts for more than 90% of the total subsidence.

Figure 8.10 illustrates the variations of the compression of each soil layer with time along Paths A1, B1, and C1. At the beginning of the building load period, the load is small and the compression of the first aquitard is larger than that of the second aquitard along Paths A1 and C1. With the load increasing, the building load is transferred to the sand layer by piles, and the compression of the second aquitard increases quickly. The building load is diffused along the vertical direction so that the second aquitard along Paths A1, B1, and C1 is compressed more than the first one. The compression of aquifers is much less than that of aquitards and is mainly caused at the loading moment.

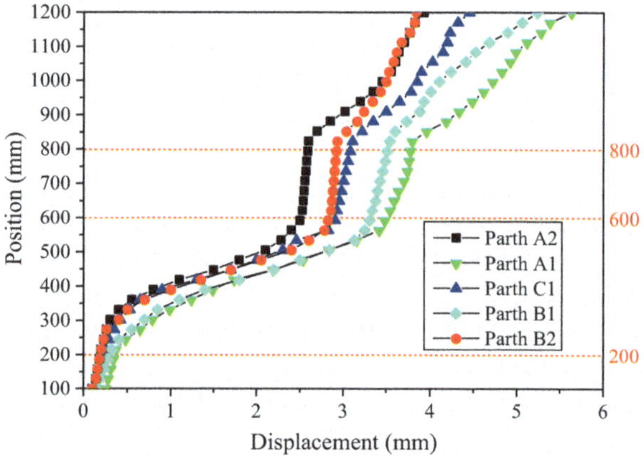

Fig. 8.9 Subsidence of soils along Paths A2, A1, C1, B1, and B2 at the building load period

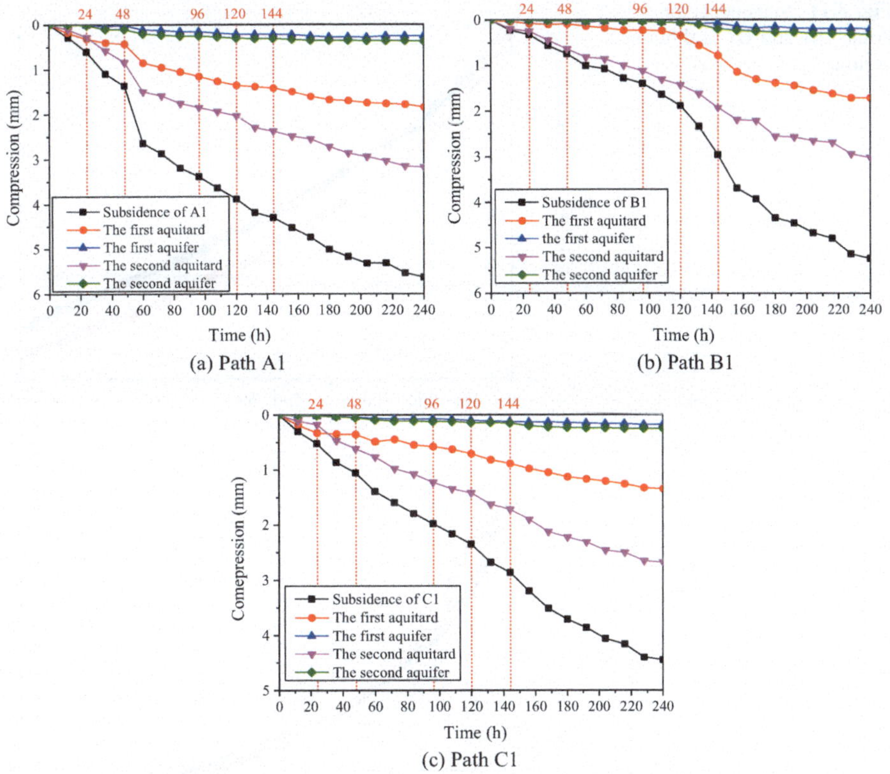

Fig. 8.10 Variations of compression of each soil layer with time along different paths

8.3.2 Subsidence at Dewatering Period

8.3.2.1 Subsidence of Buildings and the Surrounding Points

Figure 8.11 illustrates the subsidence of buildings with time during the building load and dewatering periods. The first pumping period lasts 335 h and the second pumping lasts 755 h. Building A experiences the subsidence of 3.1 and 3.8 mm in the first and second dewatering periods, while Building B experiences the subsidence of 3.6 and 3.9 mm, respectively. The subsidence of Building A is less than that of Building B at the dewatering period. The total subsidence of Building A at the loading and dewatering periods is 12.7 mm, which is close to that of Building B (12.8 mm). The subsidence of Buildings A and B is 5.8 and 5.2 mm at the building load period, accounting for 45.6 and 40.7% of the total subsidence, respectively.

The building load period lasts 240 h, while the first and second dewatering periods last 335 and 755 h, respectively. This indicates that the subsidence caused by the dewatering develops slowly but longer, compared with that caused by the

Fig. 8.11 Subsidence of Points A1 and B1 at different periods

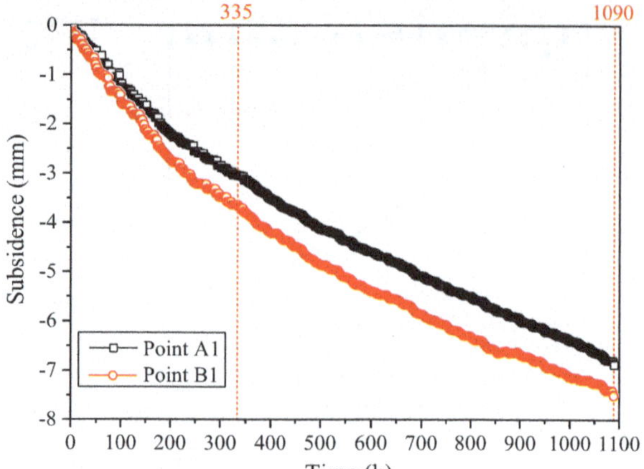

(a) Dewatering period

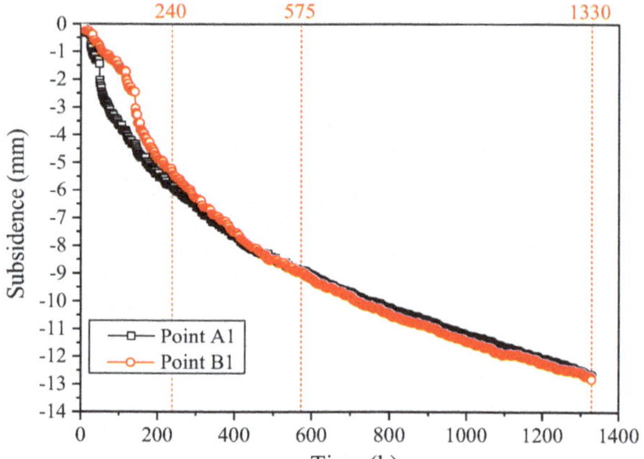

(b) Loading and dewatering period

building load. It is a remarkable fact that each pumping work only lasts 2.5 h; however, the subsidence induced by the two dewatering lasts 335 and 755 h, respectively. This means that the land subsidence caused by the dewatering is characteristic of the long duration, the slow development, and the time delay. The subsidence caused by the first pumping is close to that of the second pumping, but the latter lasts much longer. This development law is consistent with the subsidence development in Shanghai. Comprehensive groundwater exploitation program is adopted in Shanghai now. The groundwater is mainly extracted from the fourth and fifth confined aquifers and recharged to the second and third confined aquifers. The subsidence rate slows down. To avoid the final subsidence reaching the previous value, the recharging should be kept on necessarily.

The soil deformation rate does not change at the pumping moment. Although the final subsidence of Building A is close to that of Building B, the developing process is different. The subsidence of Building A is larger than that of Building B at the building load period, yet less at the dewatering period. The reason is that the soil under Building A experiences the different stress path from the soil under Building B. Soil under Building A consolidates under larger load at the beginning of loading period and under less load later while the soil under Building B is the opposite.

Figure 8.12 illustrates the variations of subsidence of building positions and the surrounding points with time. During the first dewatering period, the subsidence of Points A1, A2, B1, B2, and C1 close to the plexiglass plate is 3.1, 3.0, 3.6, 3.1, and 3.4 mm, respectively, which are nearly the same. The subsidence of Points C2 and C3 is 4.7 and 6.3 mm, respectively. The subsidence of Points C2 and C3 are larger than that of Points A1, A2, B1, B2, and C1 which are close to the plexiglass plate and are laterally restrained. Another reason is that Points C2 and C3 are closer to the pumping point. During the second dewatering period, the subsidence of Points

Fig. 8.12 Subsidence of building points and the surrounding area at different periods

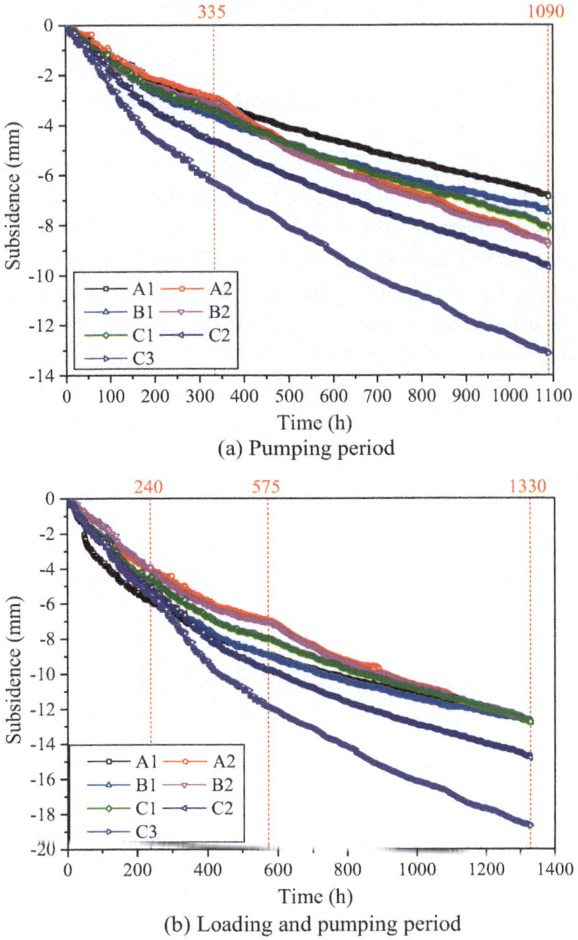

A1, A2, B1, B2, and C1 is 3.8, 5.8, 3.9, 5.7, and 4.8 mm, respectively. The subsidence of Points A1 and B1 where buildings are located is the smallest. The subsidence of Point C1 is the second place and that of Points A2 and B2 is the largest. The farther away from the buildings, the larger the subsidence. The reason is that soil stress near buildings is changed by piles when pumping. When pumping, the groundwater table drops and the additional stress in soil increases, so the soil is compressed. However, the friction force from piles prevents the soil from subsiding.

Because of the superimposition effect of buildings, the subsidence of Point C1 is larger than that of Points A2 and B2 at the building load period. When pumping, the effect of buildings on Point C1 is also more than that on Points A2 and B2 so that it subsides less. Points C2 and C3 subside 5.0 and 6.9 mm during the second dewatering period, respectively. Points C1, C2, and C3 are all 0.5 time the base width away from buildings, yet Point C3 subsides the largest at both the building load and the dewatering periods, for Point C3 is located near the pumping point and farther away from the plexiglass plate.

8.3.2.2 Distribution of Land Subsidence

Figure 8.13 shows the subsidence of points along Path M at different periods. The subsidence of Points C1–C8 is 3.4, 4.7, 6.3, 6.9, 7.2, 7.4, 8.0, and 7.8 mm at the first dewatering period and 4.8, 5.0, 6.9, 7.6, 8.0, 8.2, 8.4, and 8.9 mm at the second dewatering period, respectively. Points C4 and C8, C5 and C7 are symmetrical about the pumping point near to Point C6. Points C4 and C5 are closer to the buildings and subside less at the dewatering period. The subsidence of Point C5 is 0.8 mm less than that of Point C7 at the first dewatering period and 0.4 mm at the second dewatering period. The subsidence of Point C4 is 0.9 mm less than that of Point C8 at the first dewatering period and 1.2 mm at the second dewatering period, as shown in Fig. 8.14.

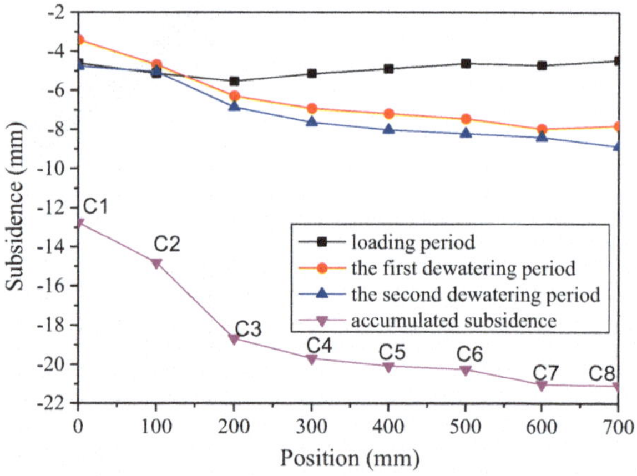

Fig. 8.13 Subsidence of points along Path M at different periods

Fig. 8.14 Subsidence of typical points at the pumping periods

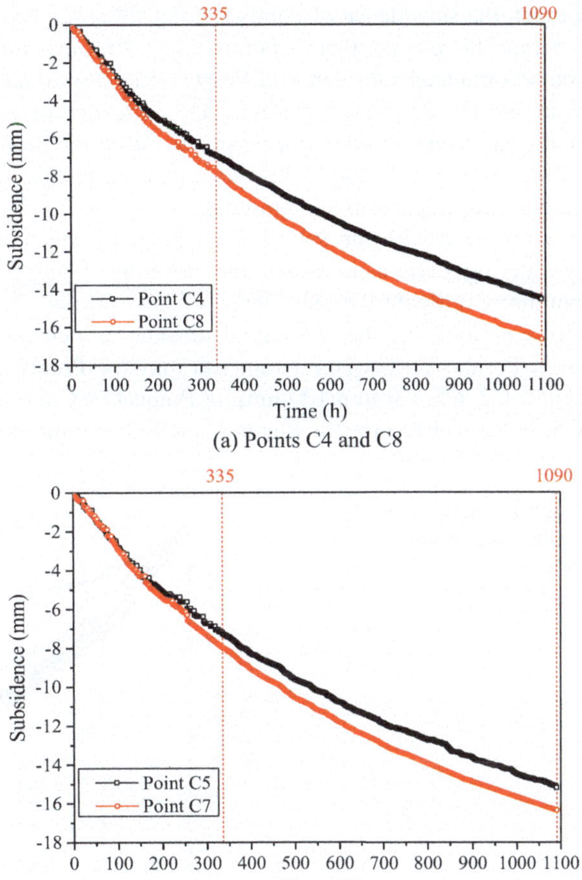

The farther the distance to the buildings, the larger the subsidence at the dewatering periods. The accumulated subsidence of Points C7 and C8, far away from the buildings, at two pumping periods, is larger than that of Points C1, C2, and C3, which are near the buildings. The plane size of the model is 1800 mm × 800 mm, which is so small that no subsidence funnel occurs, compared with the influence range of the pumping well. That is, if there are no buildings, subsidence of the soil surface should be the same at the dewatering periods. However, different area subsides differently in this physical model test. The subsidence of Points C1–C8 is 8.1, 9.7, 13.1, 14.5, 15.2, 15.6, 16.3, and 16.6 mm at two dewatering periods and 12.7, 14.8, 18.7, 19.7, 20.0, 20.2, 21.0, and 21.1 mm at the building load and two pumping periods. The subsidence of Points C7 and C8 at the dewatering periods is even larger than that of Points C1 and C2 at the building load and the dewatering periods. The reason is that Points C1 and C2 are close to the buildings, and the friction force from piles prevents them from subsiding.

In fact, the subsidence of Points C7 and C8 at the first dewatering period is 8.0 and 7.8 mm, larger than that of Points C1 (3.39 mm) and C2 (4.66 mm), but less than the accumulated subsidence of Points C1 (8.0 mm) and C2 (9.8 mm) at the building load and the dewatering periods. The influence of buildings on subsidence is not that clear at the first dewatering period. With the groundwater pumping increasing, land subsidence caused by the dewatering increases, and the negative friction effect of building piles is more significant.

Compared with the subsidence caused by the building load, the subsidence at dewatering periods is larger and develops longer, as shown in Fig. 8.15. The buildings mitigate the subsidence of area around the buildings during the dewatering periods, but the differential subsidence increases. The differential subsidence from Points C1–C8 is 2.1 mm/100 mm for Points C1–C2, 3.9 mm/100 mm for Points C2–C3, 1.0 mm/100 mm for Points C3–C4, 0.4 mm/100 mm for Points C4–C5, 0.2 mm/100 mm for Points C5–C6, 0.8 mm/100 mm for Points C6–C7, and

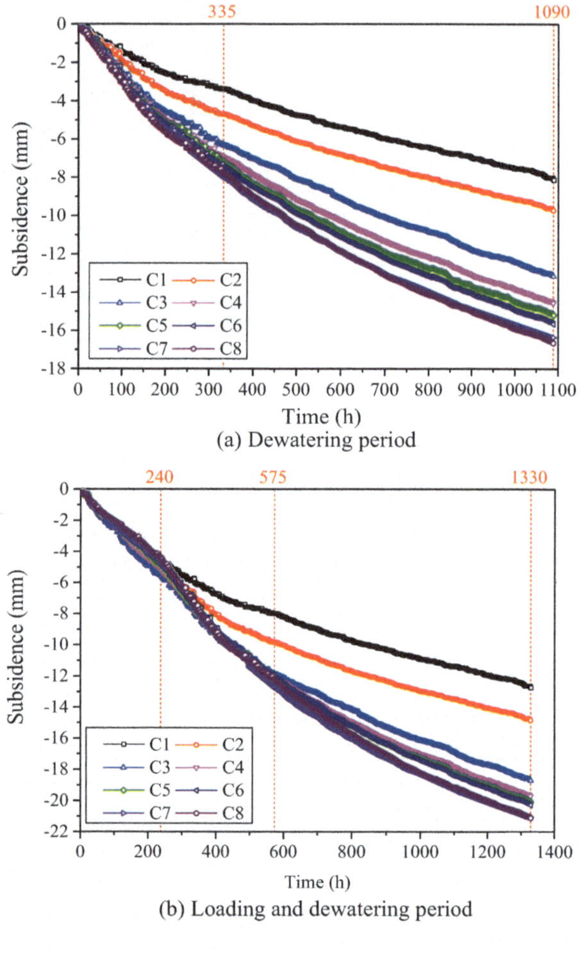

Fig. 8.15 Subsidence of points along Path M at different periods

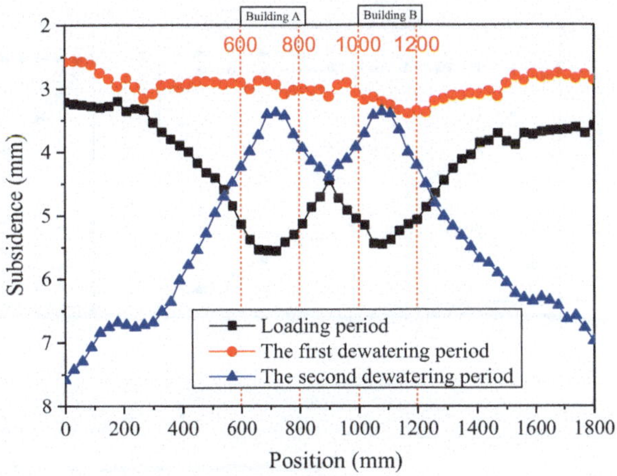

Fig. 8.16 Subsidence of points along Path L at different periods

0.1 mm/100 mm for Points C7–C8, respectively. The differential subsidence around buildings is much larger than that of the area farther away from the buildings. The subsidence caused by the dewatering is flat. The buildings increase the differential subsidence around buildings. Differential subsidence is more destructive and should be controlled carefully in the practical engineering construction.

Figure 8.16 illustrates the subsidence along Path L at different periods. The subsidence of the area close to the buildings is less than that of the area farther away from buildings significantly. The mitigation effect of buildings on subsidence during dewatering periods is dramatic. The differential subsidence at the first dewatering period is small. However, at the second dewatering period, the differential subsidence at the left of Building A and at the right of Building B is 0.8 mm/100 mm and 1.0 mm/100 mm, respectively, while that at the right of Building A and at the left of Building B is both 0.7 mm/100 mm.

8.3.2.3 Deformation of Soil Layers

When pumping, the saturation of the first aquitard decreases and the soil shrinks. Figure 8.17 illustrates the compression of each soil layer along Paths A1, B1, and C1 at the pumping periods. The compression of the two aquitards accounts for most of the land subsidence. At the first pumping period, the compression of the first aquitard is smaller than that of the second one. At the second pumping period, the unsaturated part of the first aquitard increases and the shrinkage became serious. The first aquitard experiences larger compression. For the decrease in upper soil unit weight, the aquifer rebounded about 0.3 mm. However, compared with the subsidence, the rebound can be ignored.

Compared with the subsidence caused by the dewatering with that of the building load, the former lasts much longer time. At the later stage of dewatering

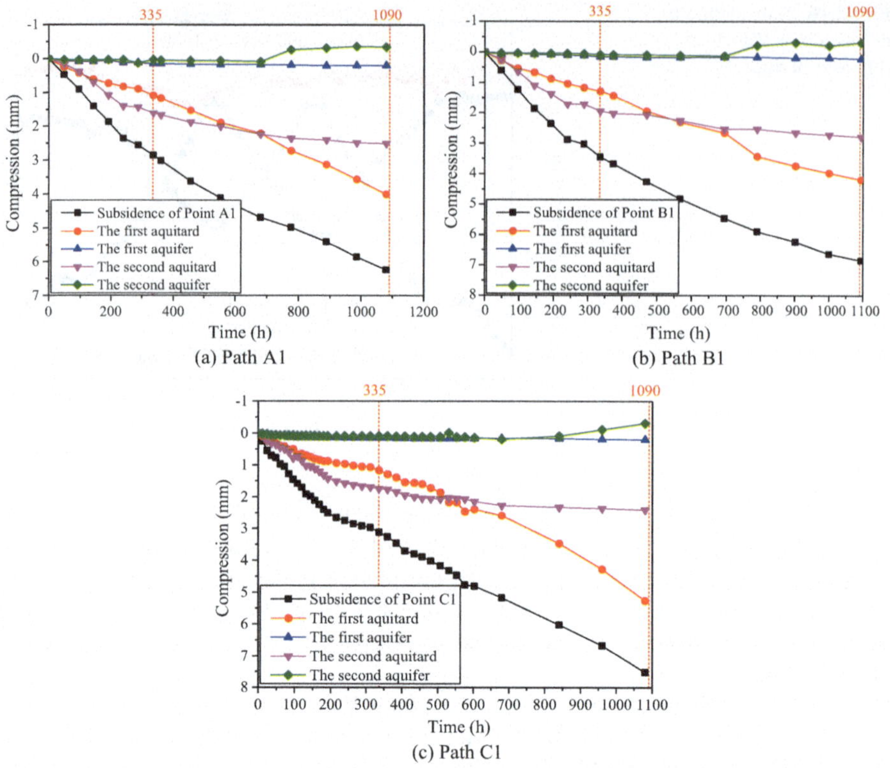

Fig. 8.17 Compression of each soil layer on typical paths at the pumping periods

period, the subsidence continues to increase because the saturation of the upper soil layer decreases and the soil shrinks. The groundwater is mainly extracted from the deeper confined aquifers and recharges to the shallower confined aquifers in Shanghai at present. The saturation of the shallower soft soil layers is invariant. Shrinkage cannot occur in the aquitard of Shanghai. In the early years, the recharge is little and the large subsidence might be partly caused by this factor. In addition, recharging to the shallow soil layers in foundation pit dewatering may be necessary to prevent soil from shrinking.

Figure 8.18 illustrates the variations of the subsidence along typical paths at different periods. At the first dewatering period, the compression of the first aquitard along Paths A2, A1, C1, B1, and B2 is 1.0, 1.4, 1.2, 1.2, and 0.9 mm, accounting for 33.6, 47.2, 37.8, 36.4, and 30.9% of the total subsidence, respectively. The compression of the second aquitard along the five paths is 2.0, 1.7, 1.7, 2.0, and 2.0 mm, accounting for 67.5, 59.4, 55.1, 60.9, and 69.4% of the total subsidence, respectively. The compression of the first aquitard is less than that of the second one. The two aquifers are compressed little. At the second dewatering period, the compression of the first aquitard along the five paths is 5.1, 2.9, 4.1, 2.9, and

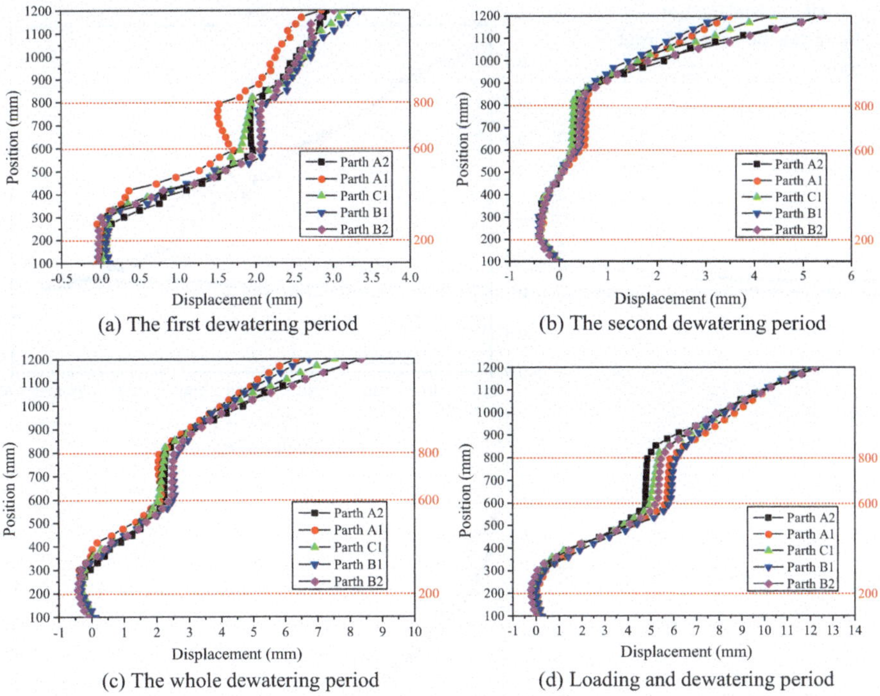

(a) The first dewatering period

(b) The second dewatering period

(c) The whole dewatering period

(d) Loading and dewatering period

Fig. 8.18 Soil subsidence along typical paths at different testing periods

4.9 mm, accounting for 93.9, 84.2, 93.2, 85.6, and 91.6% of the total subsidence. The compression of the second aquitard along the five paths is 0.6, 0.7, 0.5, 0.7, and 0.7 mm, accounting for 11.1, 20.1, 12.3, 19.4, and 12.2% of the total subsidence. The first aquitard is compressed much more than the second one at the second dewatering and the whole dewatering period. Compared with the building load period, the subsidence presents different distribution laws. At the building load and the dewatering periods, the compression of the first aquitard along the five paths is 7.4, 6.1, 6.6, 5.9, and 6.8 mm, accounting for 60.4, 51.2, 55.5, 48.9, and 55.3% of the total subsidence. The compression of the second aquitard along the five paths is 4.9, 5.5, 5.0, 5.7, and 5.4 mm, accounting for 39.8, 46.3, 41.4, 47.8, and 44.1% of the total subsidence. The dewatering changes the distribution of the subsidence developing at the building load period.

8.3.3 Subsidence at Recharging Period

Figure 8.19 shows the subsidence of buildings at the recharging period. At the end of the second dewatering period, the rates of subsidence for Buildings A and B are

Fig. 8.19 Subsidence of building positions at different periods

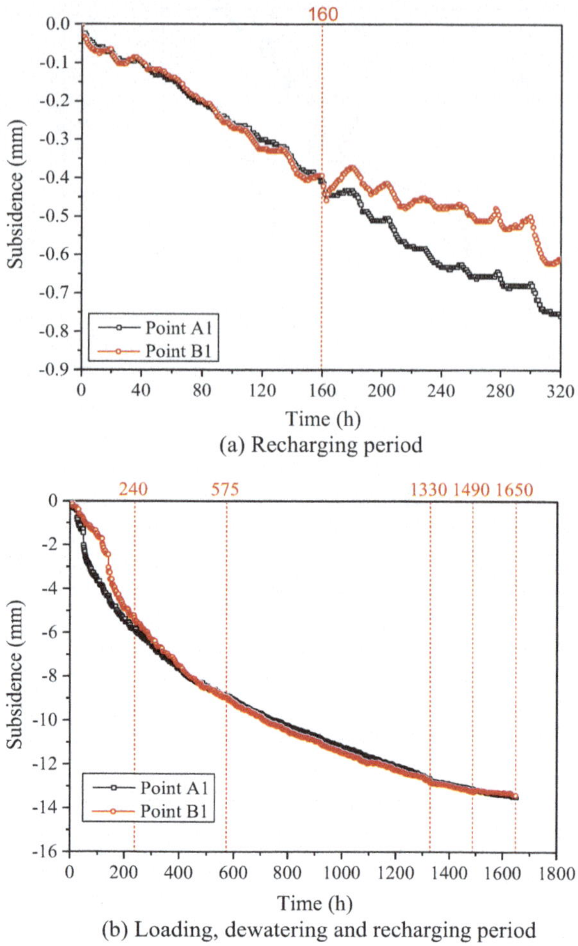

0.18 and 0.22 mm/day, respectively. Buildings A and B both subside at a rate of 0.06 mm/day at the first recharging period, while 0.05 and 0.03 mm/day at the second recharging period, respectively. At the beginning of the second recharging period, Building A rebounds about 0.01 mm, while Building B rebounds about 0.08 mm. Both buildings subside again after a moment of rebounding. The net groundwater pumping was zero at the end of the second recharging period. This indicates that the subsidence of buildings is irreversible, and the recharging can be applied to mitigate the subsidence. A certain amount of groundwater pumping is necessary for some requirements in Shanghai. To prevent land from subsiding, recharging should be kept on. It is not appropriate to estimate land subsidence by the net groundwater pumping. If the net amount of groundwater pumping is zero at certain time, it cannot be determined that land subsidence has no relationship with the groundwater pumping. The subsidence may be caused by the early pumping.

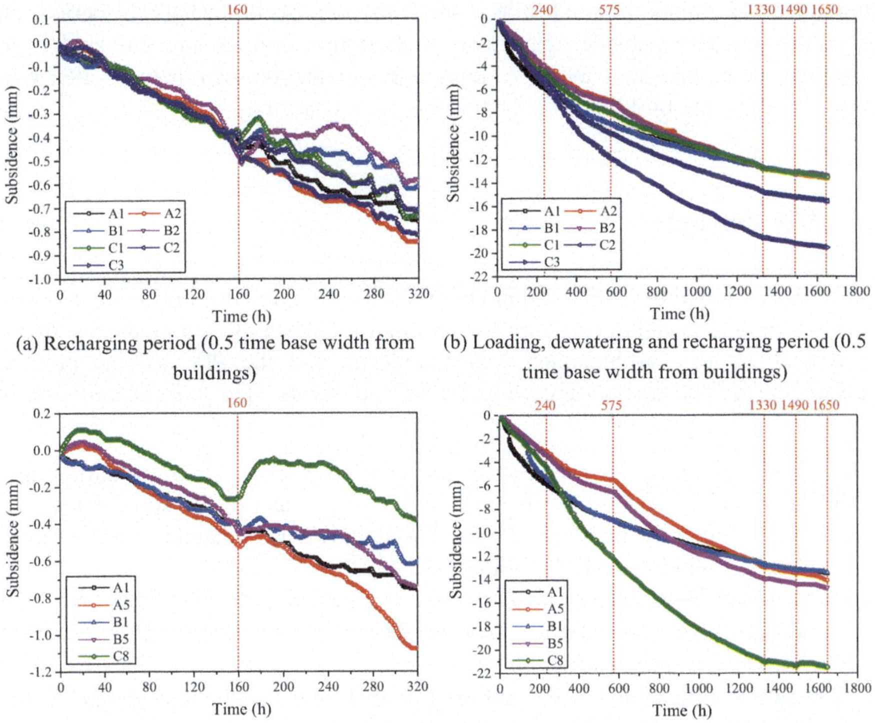

(a) Recharging period (0.5 time base width from buildings)

(b) Loading, dewatering and recharging period (0.5 time base width from buildings)

(c) Recharging period (2.5 times base width from buildings)

(d) Loading, dewatering and recharging period (2.5 times base width from buildings)

Fig. 8.20 Subsidence of building positions and the surrounding area

A large amount of groundwater is extracted at the early stage in Shanghai, and now the amount of recharging is gradually increasing. From 2011, the amount of recharging has been larger than that of pumping. However, Shanghai still subsides at a rate of about 6 mm/a. This may have a great relationship with the groundwater exploitation in the early years.

Figure 8.20 illustrates the subsidence of building positions and the surrounding area. It can be seen that recharging can help mitigate subsidence effectively but little rebound can be seen. At the end of the second dewatering period, the subsidence rates of Points A2, B2, C1, C2, and C3 which are 0.5 time the base width away from the buildings are 0.17, 0.28, 0.26, 0.22, and 0.21 mm/day, respectively. Their rates reduced to be about 0.07 mm/day at the first recharging period and 0.05 mm/day at the second recharging period. At the end of the second dewatering period, the subsidence rates of Points A5, B5, and C8 which are 2.5 times the base width away from buildings are 0.21, 0.17, and 0.27 mm/day, respectively. Their rates reduced to be 0.08, 0.06, and 0.04 mm/day at the first recharging period and 0.08, 0.05, and 0.02 mm/day at the second recharging period, respectively. Similar to the subsidence of buildings, the subsidence of the surrounding points is irreversible either.

Although some points rebound little at the beginning of the recharging period, all the points begin to subside again after a short time. So, to prevent land from subsiding, the recharging cannot be stopped. It can also be seen that subsidence of points closer to the buildings is affected less by recharging.

8.4 Conclusions

This chapter studied the deformation law of each soil layer considering dual effects of the building load and the groundwater withdrawal. The physical model test under the typical geology in Shanghai was conducted, and the PIV was adopted to measure the displacement of each point in soil layers. The conclusions are as follows:

(1) The later constructed building accelerates the subsidence rate of the earlier one, while the earlier one reduces the subsidence of the later one to some extent, for the soil foundation under the later one has already consolidated under the earlier constructed building load for some time.
(2) For the superimposition effect of land subsidence caused by buildings, the subsidence inside the building area increases while the differential subsidence outside the building area increases.
(3) The dewatering changes the subsidence distribution which develops at the building load period. When dewatering, the friction force from piles reduces the subsidence of buildings and the surrounding area, yet increases the differential subsidence.
(4) Compared with the subsidence caused by the building loads, the subsidence caused by the dewatering develops slowly and lasts a longer period of time. The compression of soft soil layers accounts for the most part of land subsidence at both the building load period and the dewatering periods.
(5) Groundwater recharge can mitigate land subsidence effectively, yet the rebound is little. It is not appropriate to estimate the land subsidence by the net groundwater pumping. If the net amount of groundwater pumping is zero at a certain time, it cannot be determined that land subsidence has no relationship with the groundwater pumping. The subsidence may be caused by the early pumping.

References

Chai JC, Shen SL, Zhu HH (2004) Land subsidence due to ground water drawdown in Shanghai. Geotechnique 54(2):143–147

Cheng S, Zhang G, Zheng RH, Sun ZY (2011) Centrifuge modeling of response of bridge due to exploiting groundwater. Rock Soil Mech 32(6):1781–1786 (in Chinese)

Cui ZD, Tang YQ, Yan XX (2010) Centrifuge modeling of land subsidence caused by the high-rise building group in the soft soil area. Environ Earth Sci 59:1819–1826

Liu HP (2010) The study on the land subsidence with the effect of high-rise buildings in Tianjin Binhai new area (in Chinese)

Shen SL, Xu YS (2011) Numerical evaluation of land subsidence induced by groundwater pumping in Shanghai. Can Geotech J 48:1378–1392

Shi XQ, Wu JC, Ye SJ, Zhang Y, XueYQ Wei ZX, Li QF, Yu J (2008) Regional land subsidence simulation in Su-Xi-Chang area and Shanghai City, China. Eng Geol 100:27–42

Shi XQ, Fang R, Wu JC, Xu HX, Sun YY, Yu J (2012) Sustainable development and utilization of groundwater resources considering land subsidence in Suzhou, China. Eng Geol 124:77–89

Sun ZY, Zhang G, Zhang JM, Li GH, Zheng RH (2008) Centrifuge modeling of ground subsidence due to groundwater pumping. China Civil Eng J 41(4):67–72 (in Chinese)

Tang YQ, Cui ZD, Wang JX (2008a) Model test study of land subsidence caused by high-rise building group in Shanghai. Bull Eng Geol Environ 67:173–179

Tang YQ, Cui ZD, Wang JX (2008b) Application of grey theory-based model to prediction of land subsidence due to engineering environment in Shanghai. Environ Geol 55:583–593

White DJ, Take WA, Bolton MD (2003) Soil deformation measurement using particle image velocimetry (PIV) and photogrammetry. Geotechnique 3(7):619–631

Wu JC, Shi XQ, Ye SJ, Xue YQ, Zhang Y, Wei ZX, Fang Z (2010) Numerical simulation of viscoelastoplastic land subsidence due to groundwater overdrafting in Shanghai, China. J Hydrol Eng 15(3):223–236

Xu YS, Ma L, Shen SL (2011) Influential factors on development of land subsidence with process of urbanization in Shanghai. Rock Soil Mech 32(S1):578–582 (in Chinese)

Xue YQ, Zhang Y, Ye SJ (2005) Land subsidence in China. Environ Geol 48:713–720

Yan XX, Gong SL, Zeng ZQ, Yu JY, Shen GP, Wang TJ (2002) Relationship between building density and land subsidence in Shanghai urban zone. Hydrogeol Eng Geol 6:21–25 (in Chinese)

Zhang AG (2005) Sustainable development and land subsidence controlling management in shanghai. Chin J Geol Hazard Control 16(1):1–4 (in Chinese)

Zhang AG, Wei ZX (2002) Past, present and future research on land subsidence in Shanghai City. Hydrogeol Eng Geol 5:72–75 (in Chinese)

Zhang WR, Duan ZL, Zeng ZQ, Shi HP (2003) Evaluation on economic losses resulted from land subsidence in Shanghai: 1921–2000. J Tongji Univ 31(6):743–748 (in Chinese)

Chapter 9
Floor Area Ratio ANFIS Model Affected by Causes of Land Subsidence

9.1 Introduction

Land subsidence is an environmental phenomenon resulting from consolidation of the ground. It is generally a relatively slow movement but may break pipelines and result in differential settlement/tilting of buildings. The causes of land subsidence include natural factors (e.g., the collapsing of loess, drainage or organic-rich soils, and karst development) and human activities (e.g., underground construction and removal of liquid and/or solids from the ground). In the coastal urban region of China, groundwater extraction is the primary cause of land subsidence (Chai et al. 2004, 2005; Gu 1998; Li et al. 2000).

In recent years, land subsidence has occurred in over 90 cities and counties in China, including Shanghai, Tianjin, Jiangsu Province, and Shanxi Province; the total area of subsidence is reported as 93,885 km^2 (Cui 2008). Shanghai was one of the first cities to experience land subsidence. In the 80 years from 1921 to the end of 2001, the total accumulative subsidence had reached 2.7 m (Zhang and Wei 2002; Xue et al. 2005; Tang et al. 2008a). Initially, this was mainly due to the irrational withdrawal of groundwater but from the 1960s, the withdrawal of groundwater was restricted and from the end of the 1970s, the pumping of groundwater was strictly controlled in the urban area of Shanghai, and the quantity of water recharged into the subsurface was always greater than that extracted by pumping such that the subsidence caused by pumping and recharging was limited in the urban area. During the 1990s, with the development of the economy, a variety of municipal works and high-rise buildings were constructed and land subsidence appeared to accelerate in Shanghai (Tang et al. 2008b; Cui et al. 2009).

In this chapter, the "geology-environmental capacity" refers to the optimum inter-relationship between the needs of human society, the economy, and the engineering structure in a certain geological area. The geology-environmental capacity is finite, e.g., there is a limit to the weight of the buildings that the ground can sustain in a certain geological situation if particular environmental criteria are to

Z.-D. Cui, *Land Subsidence Induced by the Engineering-Environmental Effect*,
https://doi.org/10.1007/978-981-10-8040-1_9

be met. In the soft soils of Shanghai, land subsidence caused by the engineering-environmental effect has resulted in great economic loss to the city, and hence this has become an important factor, restricting the scale of construction. In Shanghai, therefore, building is mainly controlled by land subsidence.

9.2 Factors Affecting Land Subsidence in Shanghai

9.2.1 Engineering Geology and Hydrogeology

Shanghai is situated on the Yangtze River delta where the alluvial deposits generally reach 250–300 m in the urban area. The upper 150 m of the alluvium is mainly of gray, soft-plastic clay and sands. The lower horizon has a wider range of colors and is mainly hard-plastic clay alternating with layers of sand and gravel (Tang et al. 2008b).

In recent years, records indicate that the main subsidence below 70 m is related to pumping/recharge, while above this the soil deformation was mainly caused by engineering works. In this upper horizon, there are three distinct thick layers of soft soil.

9.2.2 Disaster Distribution of Land Subsidence

Based on population density, engineering construction, hydraulic engineering, flood control, etc., the area of Shanghai is divided into a basically stable region and a land subsidence region which contains two subregions as shown in Table 9.1.

9.2.3 The Floor Area Ratio and Building Density

The floor area ratio is the ratio of the total floor area with an individual structure to the area of land on which the construction has taken place. The building density is the ratio of the basement area of the building to the area of land on which the construction has taken place.

As shown in Table 9.2 (Yan et al. 2002), the building areas of four typical regions, Lujiazui, Xujiahui, Zhongyuan, and Changqiao, increased from 1980 to 1999. It can be seen that the built area increased in the 1980s and 1990s relative to that in the previous decade. In most areas, the floor space more than doubled. In the same period, land subsidence increased rapidly (Table 9.3), again more than doubling between the 1980s and 1990s.

Table 9.1 Disaster distribution of land subsidence in Shanghai

Region	Subregion	Distribution range	Area/ km^2	Land subsidence	Subsidence disaster
Basic stability region		Qingpu district, Songjiang district, Jinshan district	353.34	The accumulation of land subsidence is not over 25 mm from 1980 to 1995 and the annual average subsidence is less than 1.7 mm/a	Not occurrence
Land subsidence region	Subsidence subregion	Pudong new district, Baoshan district, Minhang district, Jiading district, Fengxian district, Nanhui district, Songjiang district, Jinshan district, Qingpu district, Chongming County	5736.66	The accumulation of land subsidence is 25–100 mm from 1980 to 1995 and the annual average subsidence is 1.7–6.7 mm/a	The disaster is not obvious
	Subsidence subregion	Urban, suburb	250.50	The accumulation of land subsidence is over 100 mm from 1980 to 1995 and the annual average subsidence is over 6.7 mm/a	The disaster is more obvious

Table 9.2 Building area of four typical regions (Unit: $\times 10^6$ m^2)

Region	Before 1970s	Increase area in 1970s	Increase area in 1980s	Increase area in 1990s	Total area
Lujiazui	5.24	0.80	1.57	3.94	11.55
Xujiahui	6.80	0.72	1.75	2.30	11.57
Zhongyuan	3.47	0.51	1.57	2.78	8.33
Changqiao	2.85	0.20	1.31	2.31	6.67
Total area	18.36	2.23	6.20	11.33	38.12

Table 9.4 indicates the number of high-rise buildings constructed in the four areas together with the total floor areas. The value of the floor area ratio relates to the proportion of the number of the high-rise buildings to the number of total buildings.

Table 9.3 Land subsidence of four typical regions (Unit: mm)

Region	In 1970s	In 1980s	In 1990s
Lujiazui	−0.30	−3.52	−12.56
Xujiahui	−2.13	−3.16	−7.18
Zhongyuan	−1.37	−4.49	−15.14
Changqiao	−0.89	−2.92	−7.95

Table 9.4 The floor area ratio of four typical regions

Region	Total building area/ $\times 10^6$ $(m)^2$	Building area of high-rise building/ $\times 10^6$ $(m)^2$	Number of high-rise buildings	The ratio of high-rise buildings to total buildings/ (%)	Land area/ $\times 10^6$ $(m)^2$	The floor area ratio
Lujiazui	11.17	3.58	123	32.01	7.37	1.53
Xujiahui	9.79	2.33	126	23.77	6.98	1.40
Zhongyuan	8.09	0.57	47	57.25	6.88	1.18
Changqiao	6.17	0.42	36	42.34	6.89	0.90

9.2.4 Dewatering and Utilization of Groundwater

Shanghai has more than one hundred years of groundwater mining history, with the national economic development of groundwater demand, the groundwater exploitation of each aquifer increased year by year. In 1960s, the groundwater was mainly exploited in the second aquifer in the urban area. From 1970s, in the suburbs, the groundwater exploitation from the second, third, fourth, and five aquifers increased year by year. Before 1965, the quantity of groundwater exploitation in Shanghai reached 135–203 million m^3/a, which was concentrated in the urban area where experienced serious land subsidence. From 1965, the quantity of groundwater exploitation decreased and it was relatively stable at the end of 1970s, which was 58–116 million m^3/a. At the same time, the recharge of groundwater was conducted in the second and third aquifers, which reached 4–18 million m^3/a. Since 1980s, the water demand increased sharply. The level and region of groundwater exploitation changed. In the urban area, the second and third aquifers were mainly exploited and the amount of the exploitation is less than that of the recharge. However, in the suburbs, the groundwater was mainly exploited from the fourth and fifth aquifers. The amount of groundwater exploitation was 101–151 million m^3/a. The amount of recharge in the suburbs was also concentrated in the second and third aquifers, only a small amount of recharge occurring in the fourth and fifth aquifers. The total recharge reached 15–26 million m^3/a. The groundwater exploitation was concentrated from June to September every year, which accounted for about 80% of the total amount of annual groundwater exploitation. The artificial recharge was mainly conducted from December to April, which accounted for over 80% of the annual recharge. Therefore, the exploitation and the recharge of groundwater in Shanghai had the characteristics of the regional concentration, the level concentration, and the time concentration.

9.3 ANFIS Model Based on the Floor Area Ratio

The adaptive neuro-fuzzy inference system (ANFIS) combines artificial neural networks (Mamdani and Assilian 1975; Jacobs 1988; Hoke and Brown 1997) with the (Sugeno) fuzzy logic inference. The typical fuzzy logic rule is given in Appendix B.

Land subsidence *S*, the building density *D*, the geological disaster district *C*, the exploitation of groundwater *K*, the recirculation of groundwater *H*, and the type of the geology structure *J* were selected as the variables to infer the floor area ratio *R* in this paper, that is, $X = (S,\ D,\ C,\ K,\ H,\ J)$. The relationship between *R* and *X* is

$$R = F(\{X\})$$

This can be established using ANFIS as follows.

(1) The finite groups of the training data are obtained at different places according to the influencing factor *X* and the floor area ratio *R*.
(2) *X* and *R* are taken as the input data and the output data of the initial structure of ANFIS, respectively. After training, the optimum structure model of ANFIS reflecting the global mapping relationship is built up between *X* and *R*.
(3) Putting the data of four typical regions into this structure model of ANFIS, the output is the predicted optimum value of the floor area ratio.

Nineteen examples were selected as the training data, from different places in Shanghai; the variables are shown in Table 9.5. The variables of the 16 examples in Table 9.5 were treated in the standardized way. The *S*, *D*, *C*, *K*, *H*, and *J* are the input of the network and the *R* is the output of the network. A model structure was built up by the subtraction cluster method. The values of the training parameters were range of influence 0.27, squash factor 1.25, accept ratio 0.5, and reject ratio 0.15.

Every input variable of the model was automatically endowed with 16 Gauss membership functions. The model structure based on the floor area ratio is shown in Fig. 9.1. Using the hybrid learning algorithm, the network was trained by the selected data. The training degree was 1000 times and the smallest root-mean-square deviation of the training data was 0.0030208. The change of the error in the training process is shown in Fig. 9.2.

The membership functions of land subsidence *S*, the building density *D*, the geological disaster district *C*, the exploitation of groundwater *K*, the recirculation of groundwater *H*, and the type of the geology structure *J* can be obtained before and

Table 9.5 ANFIS training data

	S	*D*	*C*	*K*	*H*	*J*	*R*
1	150	0.296	100	3166.7650	184.3086	100	7.86
2	100	0.301	70	4168.9943	240.2756	50	5.62
3	60	0.260	40	4562.6321	216.7324	50	1.5
4	85	0.274	70	3213.4352	192.1249	50	2.0
5	90	0.267	70	4316.2984	209.3765	50	2.0
6	120	0.288	100	3200.2712	196.0024	100	4.78
7	95	0.320	100	457.8083	763.2687	60	3.90
8	60	0.315	100	586.4392	999.9386	60	2.0
9	120	0.302	100	578.4218	999.7294	60	1.5
10	140	0.321	100	573.2672	889.8367	60	2.0
11	105	0.330	100	590.3024	1001.2626	60	1.5
12	30	0.320	100	19.1032	16.2270	80	1.5
13	40	0.332	100	22.6857	30.7264	80	2.0
14	35	0.331	100	27.6858	33.0413	80	2.0
15	25	0.317	100	28.1023	32.7869	80	1.5
16	30	0.343	100	26.7474	32.976	80	4.3
17	40	0.284	40	1709.3189	10.8912	50	1.0
18	50	0.293	40	1379.2179	5.1610	50	1.5
19	45	0.302	40	936.3198	3.0826	50	2.0

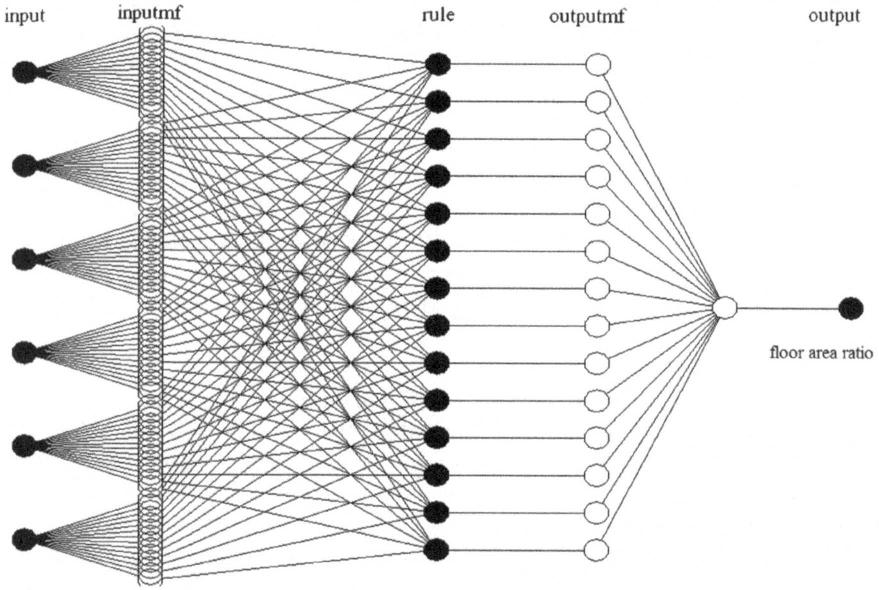

Fig. 9.1 The structure of ANFIS

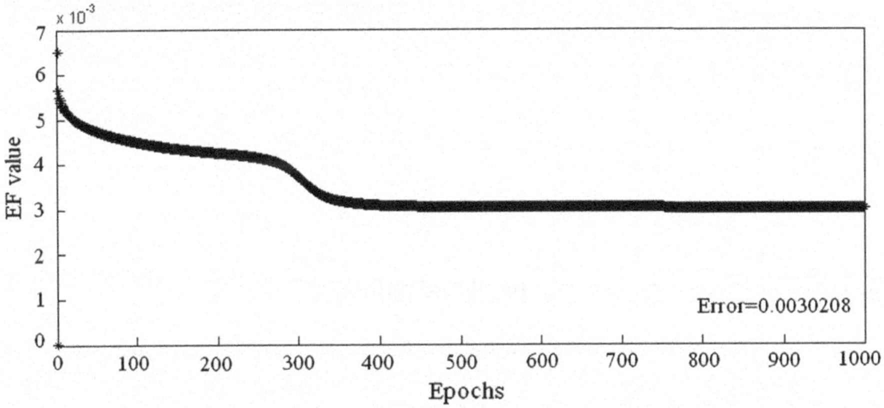

Fig. 9.2 The training error in the training process

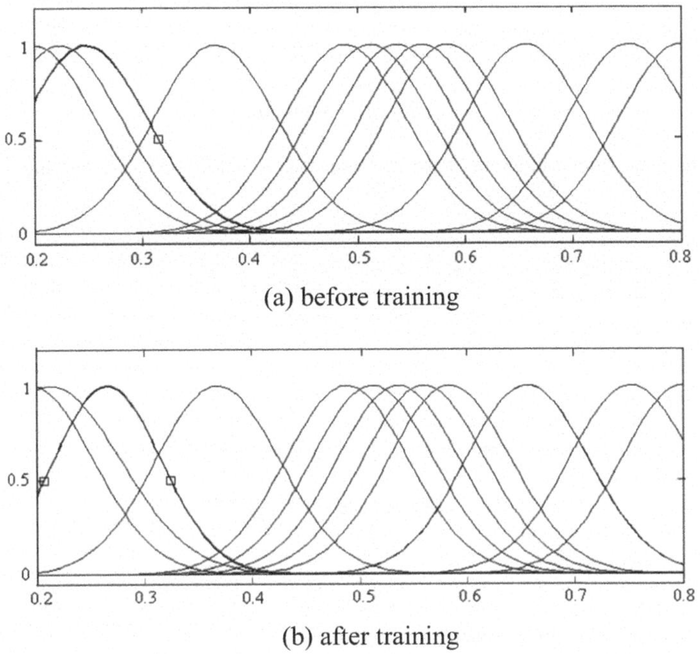

Fig. 9.3 The membership function of land subsidence before and after training

after training. The membership functions of land subsidence, the building density, the geological disaster district, the exploitation of groundwater, the recirculation of groundwater, and the type of the geology structure are illustrated in Figs. 9.3, 9.4, 9.5, 9.6, 9.7, and 9.8, respectively.

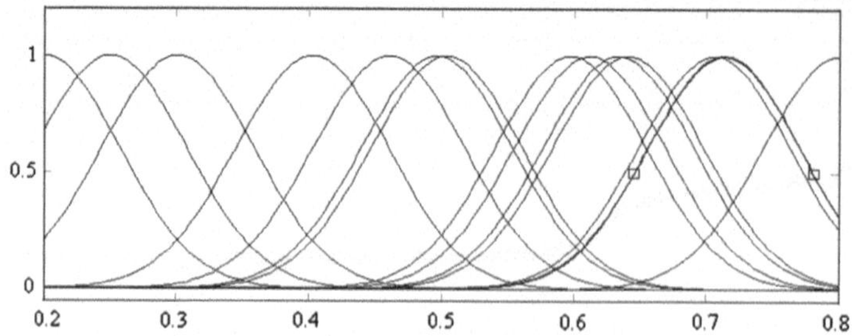

(a) before training

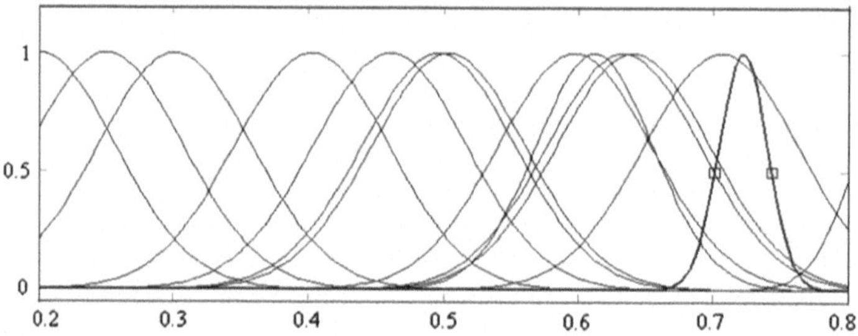

(b) after training

Fig. 9.4 The membership function of the building density before and after training

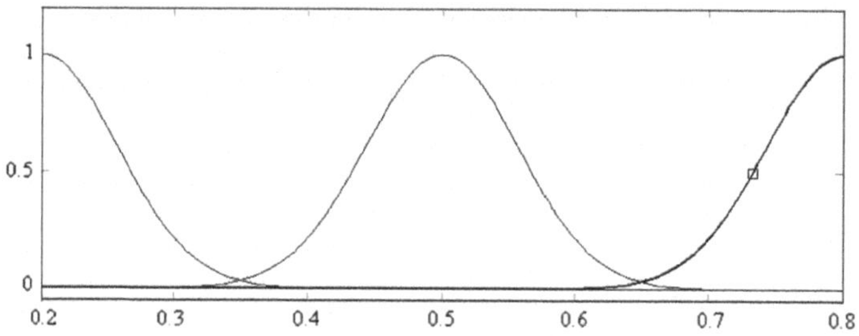

(a) before training

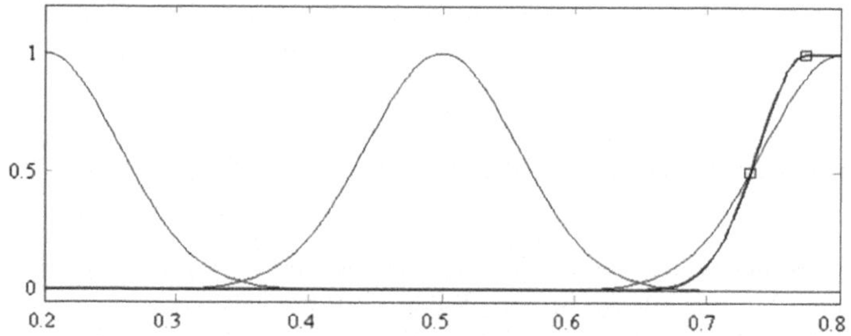

(b) after training

Fig. 9.5 The membership function of the geological disaster district before and after training

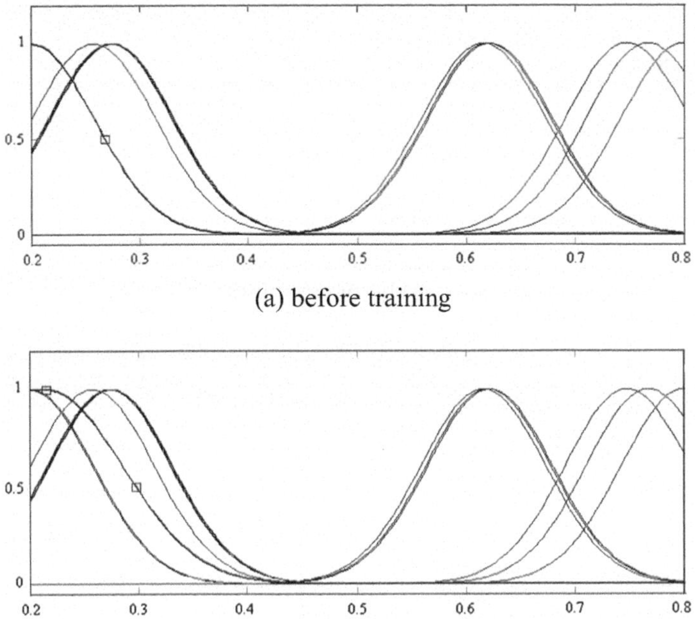

(a) before training

(b) after training

Fig. 9.6 The membership function of the exploitation of groundwater before and after training

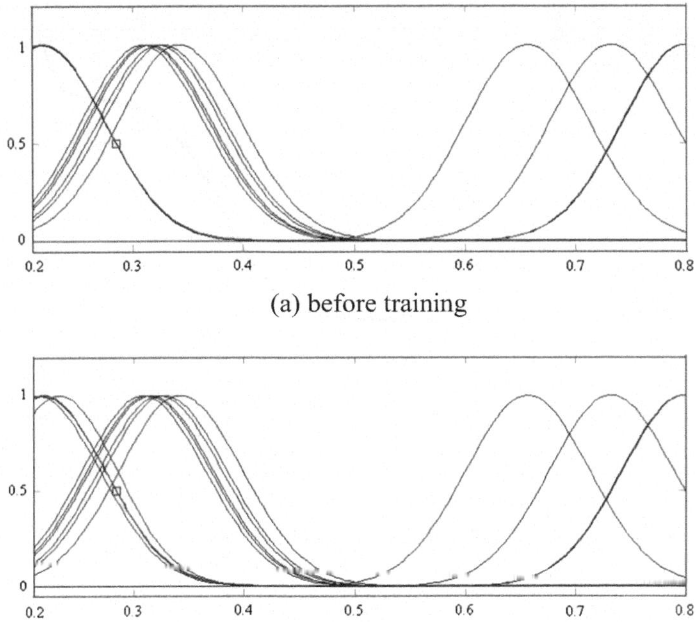

(a) before training

(b) after training

Fig. 9.7 The membership function of the recirculation of groundwater before and after training

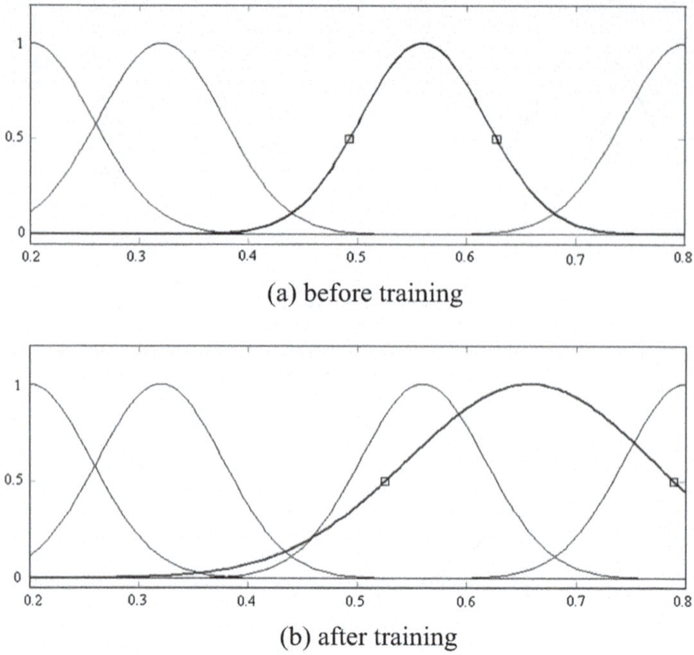

Fig. 9.8 The membership function of the type of the geology structure before and after training

Fig. 9.9 The changing law between *S* and *R*

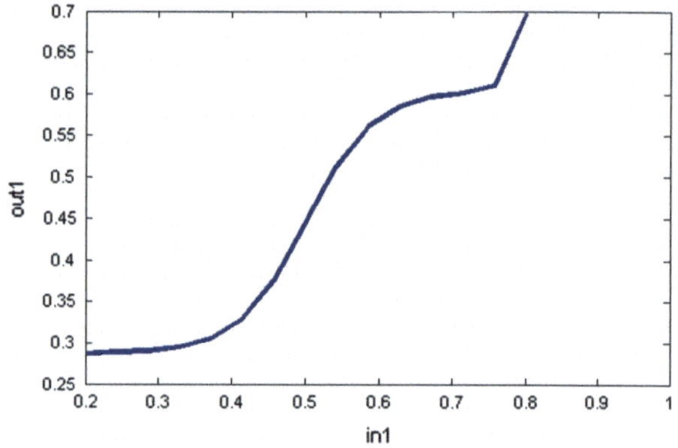

Six pictures of the changing laws between one single-input parameter of *S*, *D*, *C*, *K*, *H*, *J*, and the output parameter *R* can be obtained. The changing laws of *S*-*R*, *D*-*R*, *C*-*R*, *K*-*R*, *H*-*R*, and *J*-*R* are illustrated in Figs. 9.9, 9.10, 9.11, 9.12, 9.13, and 9.14, respectively.

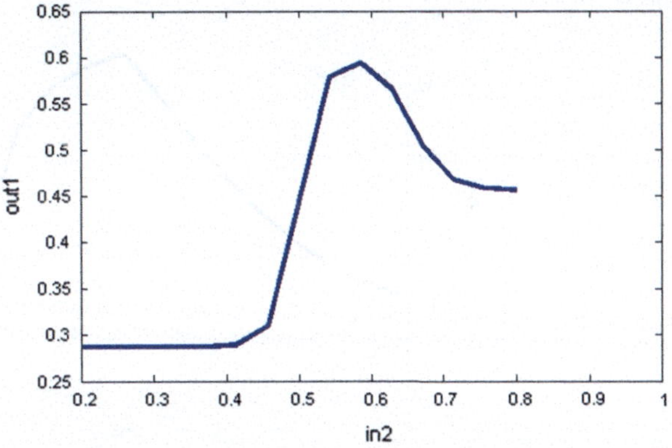

Fig. 9.10 The changing law between D and R

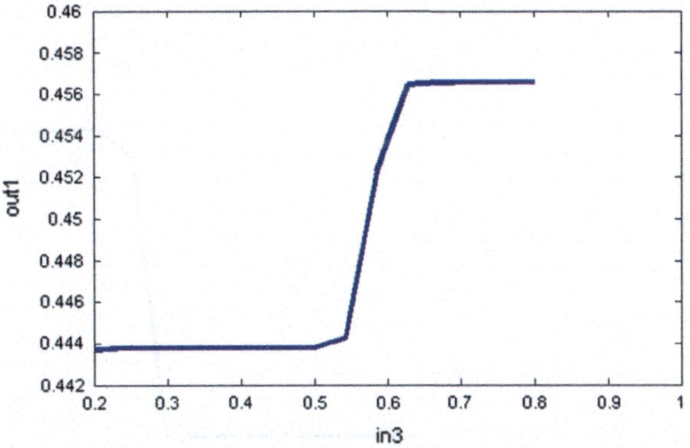

Fig. 9.11 The changing law between C and R

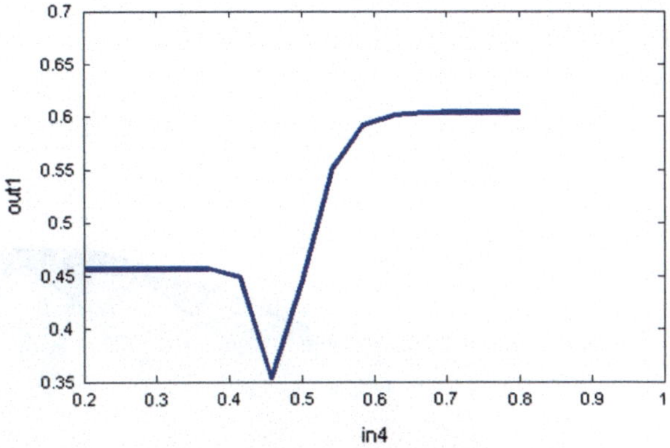

Fig. 9.12 The changing law between K and R

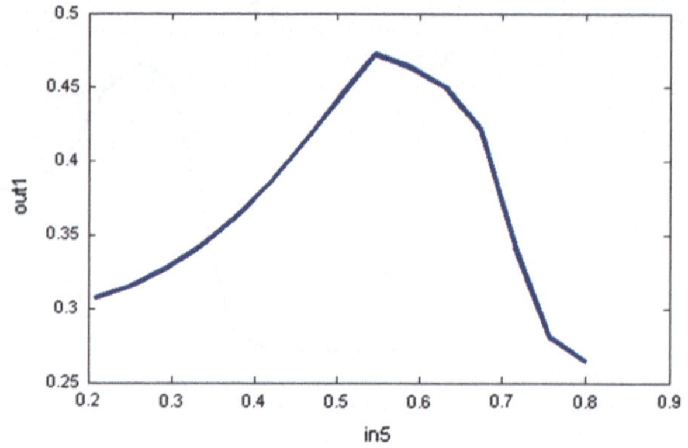

Fig. 9.13 The changing law between *H* and *R*

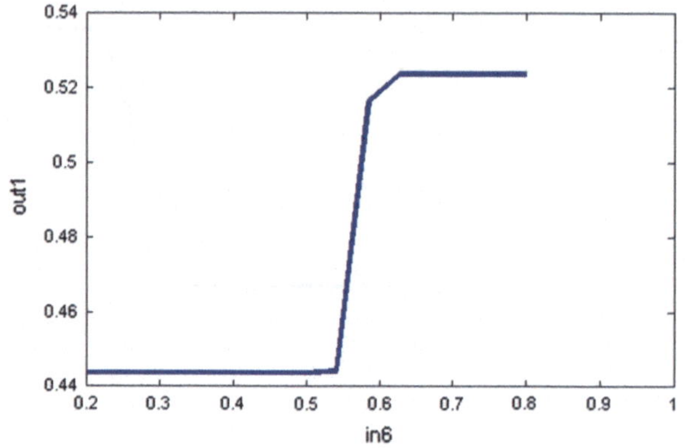

Fig. 9.14 The changing law between *J* and *R*

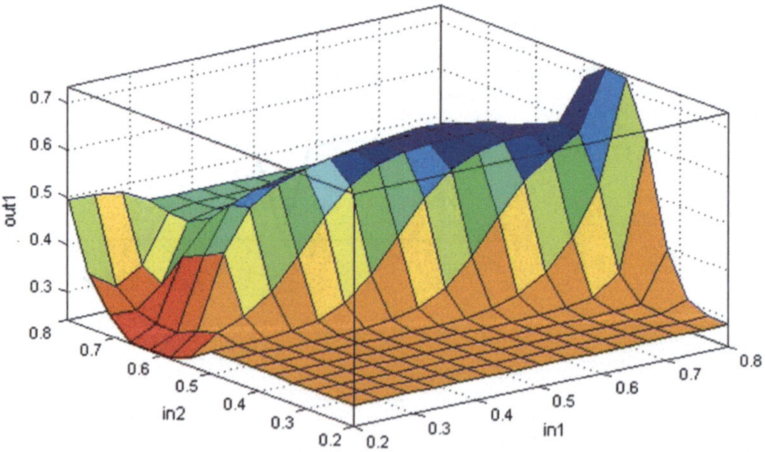

Fig. 9.15 The changing law between *S*, *D*, and *R*

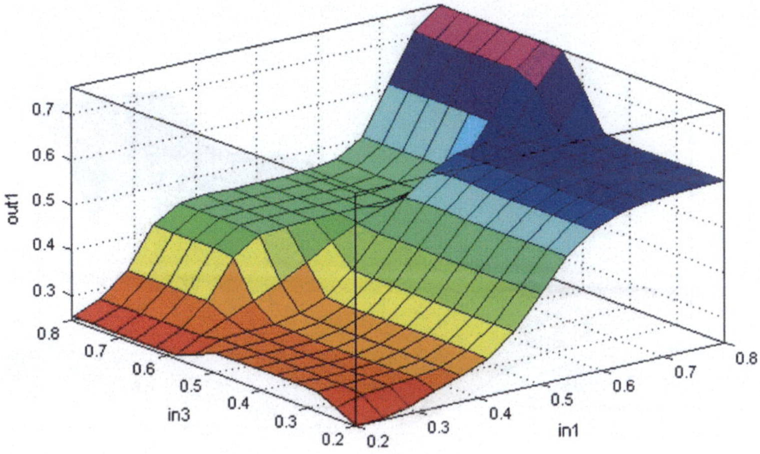

Fig. 9.16 The changing law between *S*, *C*, and *R*

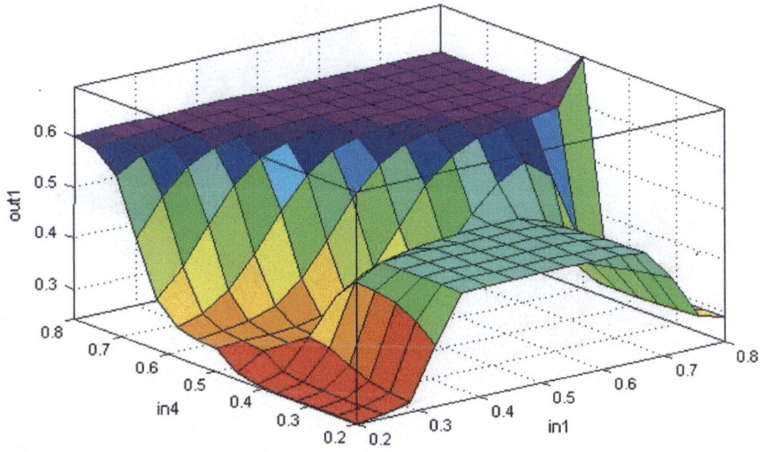

Fig. 9.17 The changing law between *S*, *K*, and *R*

Fifteen pictures of the changing laws between every two input parameters of *S*, *D*, C, *K*, *H*, *J*, and the output parameter *R* can also be obtained. The changing laws of *S*, *D-R*, and *S*, *C-R* as two examples, are shown in Figs. 9.15, 9.16, 9.17, 9.18, 9.19, 9.20, 9.21, 9.22, 9.23, 9.24, 9.25, 9.26, 9.27, 9.28, and 9.29, respectively.

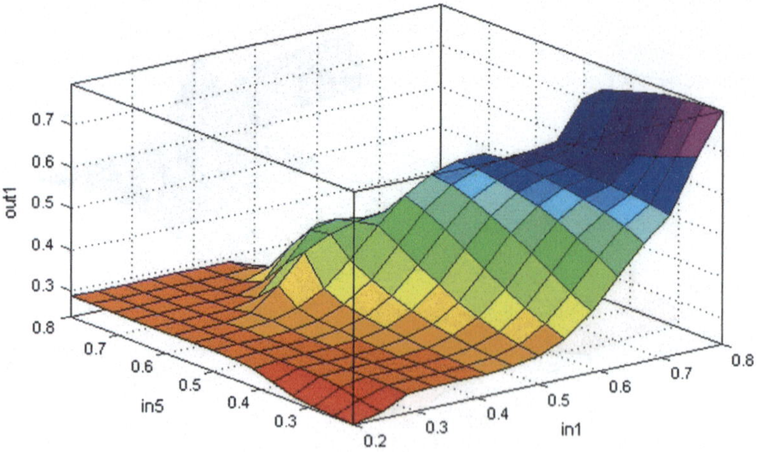

Fig. 9.18 The changing law between *S*, *H*, and *R*

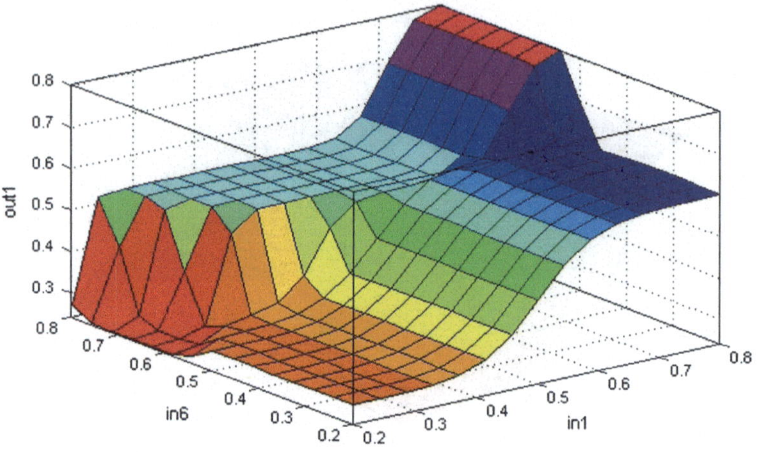

Fig. 9.19 The changing law between *S*, *J*, and *R*

9.4 The Floor Area Ratios of Four Typical Regions

Evaluation of the geology-environmental capacity of the buildings based on the ANFIS model of the floor area ratios is shown in Figs. 9.30, 9.31, 9.32, and 9.33. Whereas the floor area ratios of the regions are bigger, the high-rise buildings are

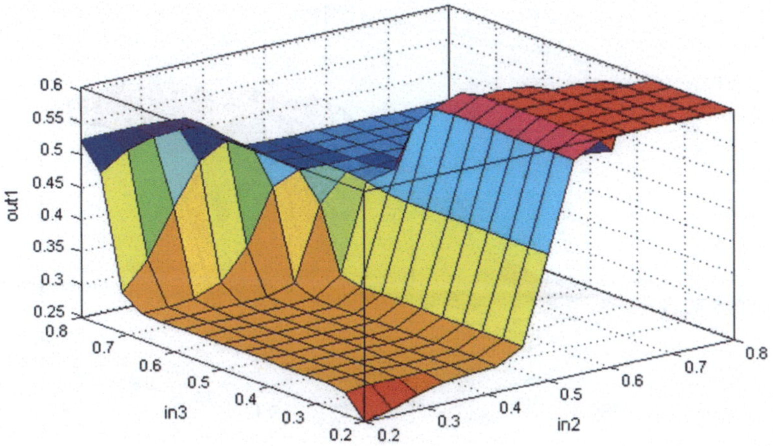

Fig. 9.20 The changing law between *D*, *C*, and *R*

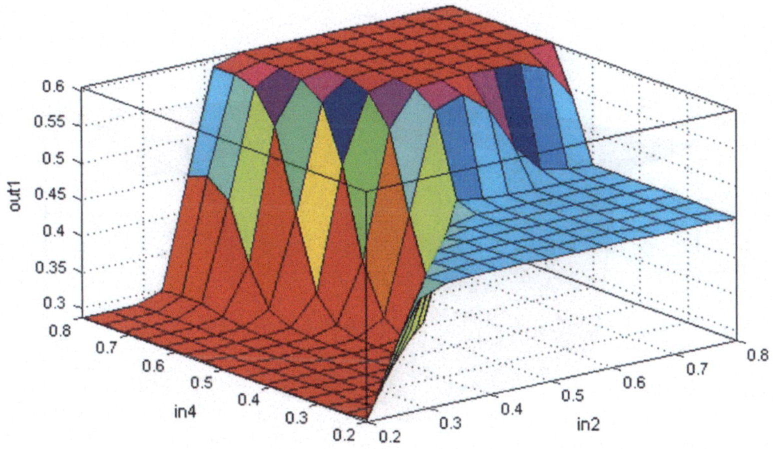

Fig. 9.21 The changing law between *D*, *K*, and *R*

dense and land subsidence is larger. It is inappropriate to continue to plan high-rise construction in these regions. However, whereas the floor area ratios of the regions are smaller and land subsidence is less, high-rise buildings can be constructed.

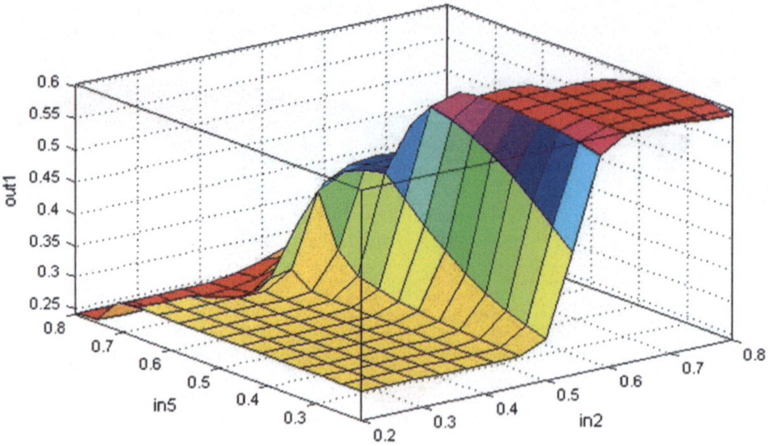

Fig. 9.22 The changing law between *D*, *H*, and *R*

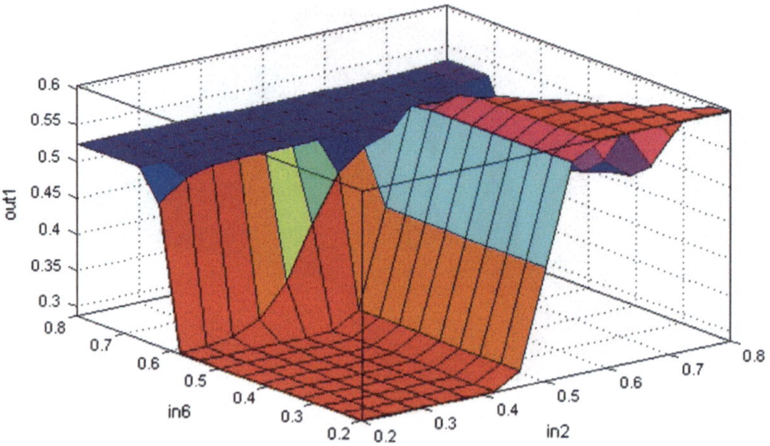

Fig. 9.23 The changing law between *D*, *J*, and *R*

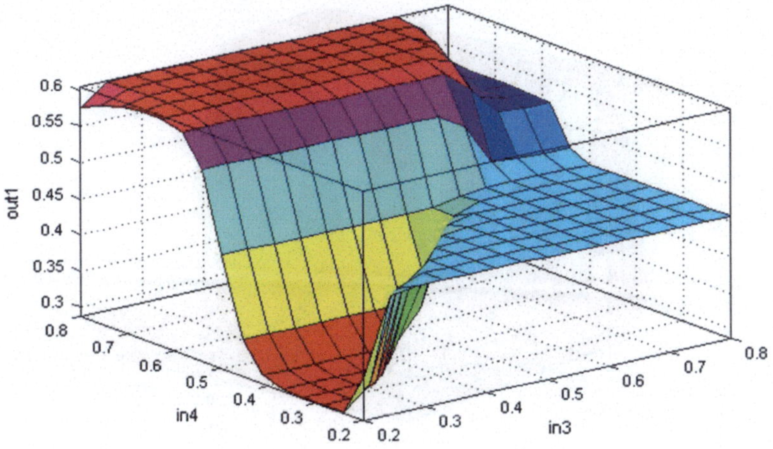

Fig. 9.24 The changing law between *C*, *K*, and *R*

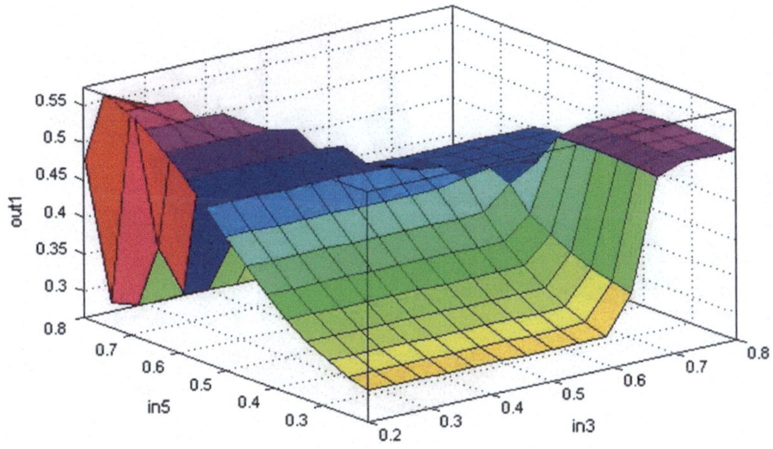

Fig. 9.25 The changing law between *C*, *H*, and *R*

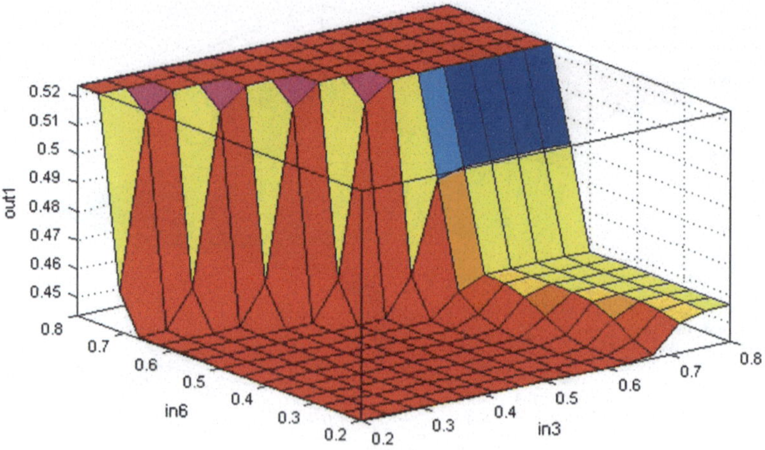

Fig. 9.26 The changing law between *C*, *J*, and *R*

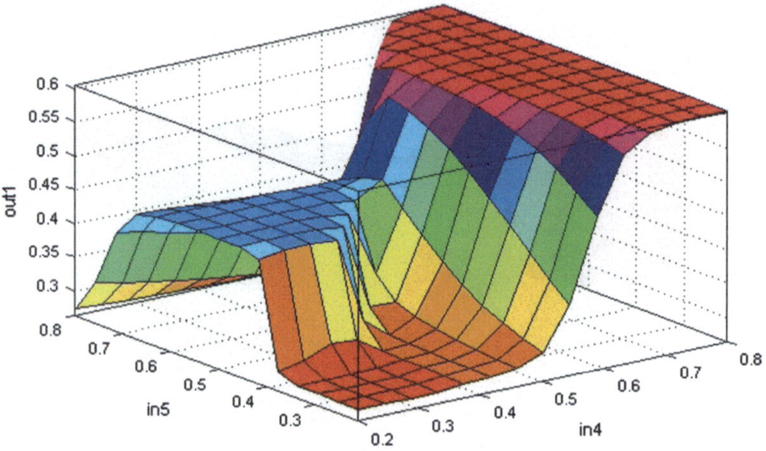

Fig. 9.27 The changing law between *K*, *H*, and *R*

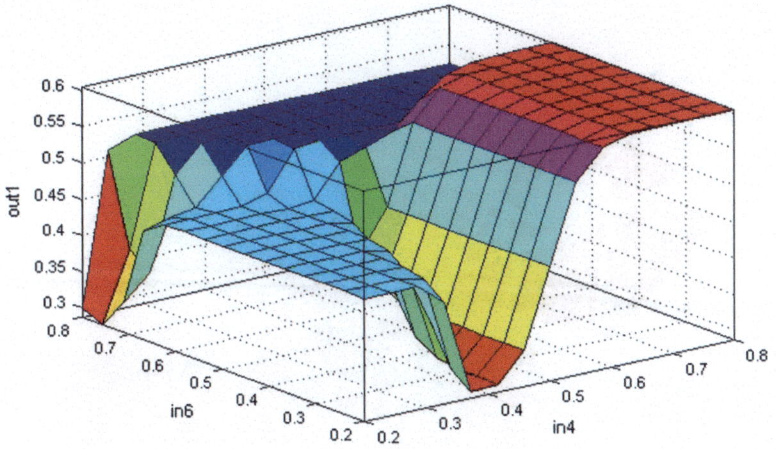

Fig. 9.28 The changing law between *K*, *J*, and *R*

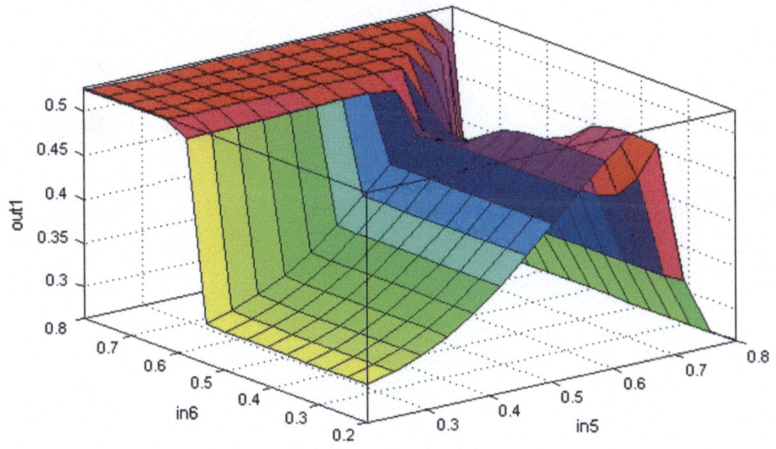

Fig. 9.29 The changing law between *H*, *J*, and *R*

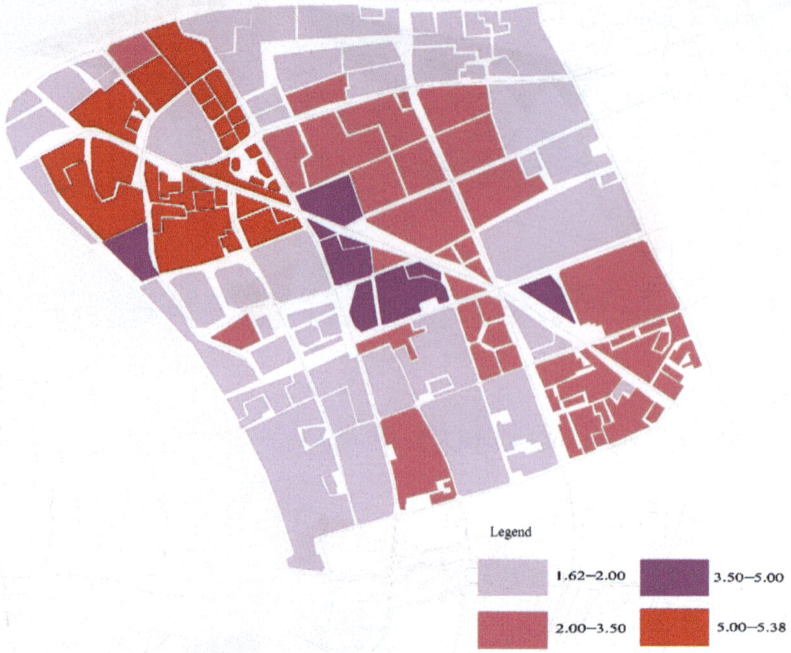

Fig. 9.30 The floor area ratios of the first typical region

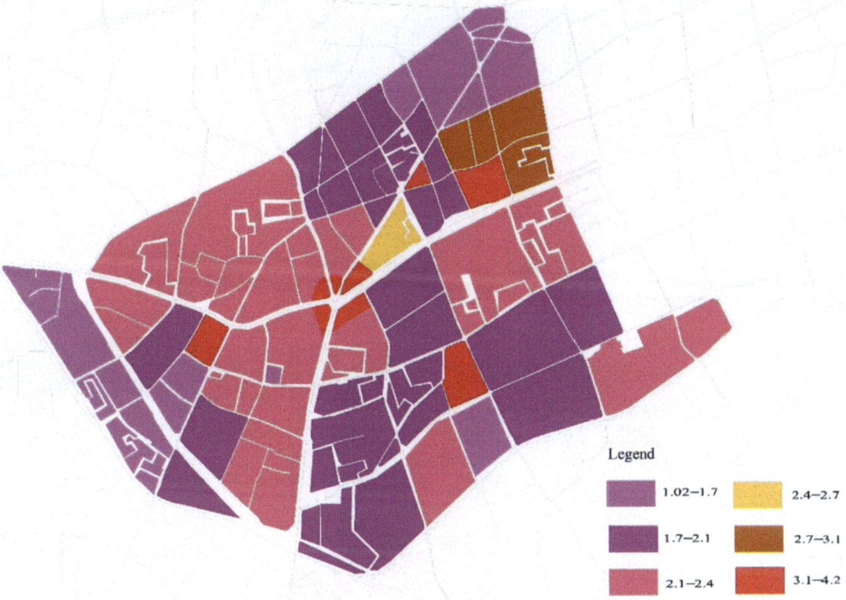

Fig. 9.31 The floor area ratios of the second typical region

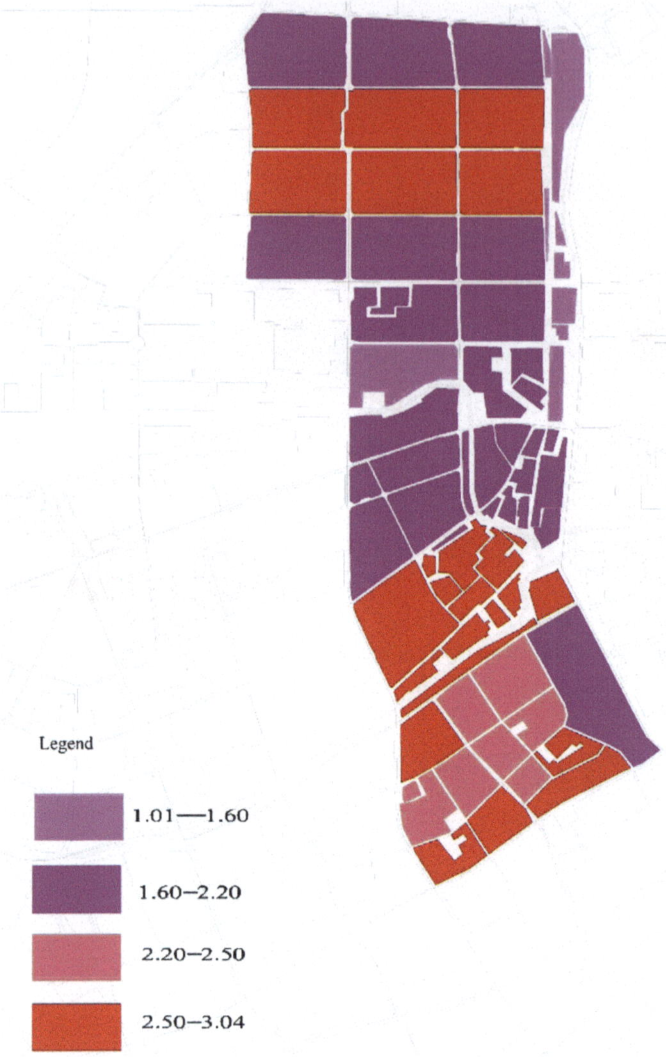

Fig. 9.32 The floor area ratios of the third typical region

Fig. 9.33 The floor area ratios of the fourth typical region

9.5 Conclusions

This chapter has demonstrated that the geological environmental capacity of the building is mainly controlled by the land subsidence and the relationship can be assessed using the floor area ratio. ANFIS was used to evaluate the floor area ratios of four typical areas in the Shanghai region in order to offer some guidance in respect of urban planning.

References

Chai JC et al (2004) Land subsidence due to groundwater drawdown in Shanghai. Geotechnique 54(3):143–148

Chai JC et al (2005) 1D analysis of land subsidence in Shanghai. Lowland Technol Int 7(1):33–41

Cui ZD (2008) Study of land subsidence caused by the dense high-rise building group in the soft soil area (in Chinese). Tongji University, Shanghai, China

Cui ZD, Tang YQ, Yan XX (2009) Centrifuge modeling of land subsidence caused by the high-rise building group in the soft soil area. Environ Earth Sci. https://doi.org/10.1007/s12665-009-0163-9

Gu XY (1998) Review and prospects of land subsidence computation (in Chinese). Chinese J Geolog Hazard Control 93(2):81–85

Hoke E, Brown ET (1997) Practical estimates of rock mass strength. Int J Rock Mech Min Sci Gabstr 34(8):106–119

Jacobs RA (1988) Increased rates of convergence through learning rate adaptation. Neural Networks 17(1):295–307
Li QF, Fang Z, Wang HM (2000) A mathematical model and forecast of groundwater workable reserves for Shanghai (in Chinese) Shanghai. Geology 23(2):36–43
Mamdani EH, Assilian S (1975) An experiment in linguistic synthesis with a fuzzy logic controller. Int J Man-Machine Stud 7(1):1–13
Tang YQ et al (2008a) Application of grey theory-based model to prediction of land subsidence due to engineering environment in Shanghai. Environ Geol 55(3):583–593
Tang YQ et al (2008b) Model test study of land subsidence caused by the high-rise building group. Bull Eng Geol Env 67(2):173–179
Xue Y et al (2005) Land subsidence in China. Environ Geol 48:713–720
Yan XX et al (2002) Relationship between building density and land subsidence in Shanghai urban zone (in Chinese). Hydrogeol Eng Geol 33(6):21–26
Zhang AG, Wei ZX (2002) Past, present and future research on land subsidence in Shanghai (in Chinese). Hydrogeol Eng Geol 33(5):72–75

Chapter 10
Conclusions and Prospects

10.1 Conclusions

In this monograph, the engineering-environmental effect refers to the high-rise building load and the dewatering of foundation pit for the construction of underground structures and the land subsidence caused by the engineering-environmental effect is studied. Land subsidence exists in the cities throughout the world. With the increase of the high-rise buildings, land subsidence induced by the human factors accelerates. Though land subsidence occurs slowly, the accumulation results in the crack of the road, the damage of the pipelines, the tilt or the crack of the buildings, and so on. It causes loss of economy. With a large number of high-rise buildings being built and the dewatering of foundation pit for construction of underground structures, the engineering-environmental effect becomes the main cause of land subsidence. In this monograph, the mechanism of land subsidence and the floor area ratio are studied by the theoretical analysis, the in-site monitoring, the general model test, and the centrifugal model test combined with the scanning electron microscope test (SEM) and the mercury intrusion porosimetry test (MIP). The results are as follows:

(1) The elastic modulus of overlying soil has little influence on ground surface displacement caused by decompression of confined water, yet thickness of overlying soil should be taken into consideration. Settlement of soils overlying confined aquifer increases up to down nearby the pumping well. Over a certain distance, the settlement does not change with buried depth. For unsteady flow confined wells, the settlement value and region at ground and aquifer surface grow with pumping time increasing. The maximum horizontal displacement value and position at ground surface do not change with time after pumping for a certain time. The length of dewatering well should be considered in the consolidation analysis of layered soils caused by water pumping from the well. The increase in the length of the well leads to smaller surface displacement. The larger values of anisotropic permeability parameters can lead to the smaller

Z.-D. Cui, *Land Subsidence Induced by the Engineering-Environmental Effect*,
https://doi.org/10.1007/978-981-10-8040-1_10

surface displacements and excess pore pressure. The layered soil cannot be regarded easily as the uniform soil by taking a weighted average of soil parameters, since this method leads to great deviation of the description of consolidation behaviors of layered soils. Pumping from the shallow aquifer can result in more serious differential settlement in the surrounding areas of centerline, while pumping from the deep aquifer can lead to the larger influence range of settlement. The increase in the recharging ratio can effectively restrict the vertical surface settlement and horizontal surface displacement. This effect is particularly obvious when the amounts of pumping and recharging reach the balance. The displacements induced by the cyclic pumping and recharging are smaller than those caused by pumping alone, although their pumping amplitudes are the same. The change of surface displacements induced by pumping lags behind that of the excess pore pressure.

(2) Building itself may experience a maximum settlement because of the remarkable subsidence superimposition effect between the high-rise buildings which exceeds the allowable values. However, the land subsidence decreases dramatically with the distance increasing. The range of the land subsidence caused by the building load is about 300 m from the in-site monitoring. The mucky clay of layer No. 4 has the maximum settlement than other soil layers. A slight rebound occurs in the silt sand of layers No. 7 and No. 9 because a large amount of groundwater recharge has been conducted in Shanghai in recent years. As a permeable material, particles of the sand bear less effective stress as the groundwater rises, causing a rebound in the sand layer. The surrounding ground of high-rise buildings suffers obviously superimposed subsidence by the additional stress, which causes a larger ground settlement than the estimated value. Therefore, in the design stage, it is reasonable to control the distance between the dense high-rise buildings in order to avoid the settlement of regional ground over the allowable value and affect the normal use of other structures.

(3) In the centrifuge model tests, the silty clay of layer No. 4 experiences the maximum subsidence. The exponential function of three-order attenuation can well fit the subsidence of different soil layers with time. It can predict the subsidence of soil layers at any time. The central area of the building group has larger subsidence and the subsidence superimposition effect is obvious. It can exceed the allowance and cause land subsidence hazard. The land subsidence affected by the different building distances is studied by the centrifuge model tests. The building distance is smaller; the subsidence superimposition effect is more obvious. So the building distance among the building group can be properly increased to decrease the subsidence superimposition effect on the area. In the centrifuge model tests, the larger excess pore water pressure in the bearing stratum for the pile tip of the central area also shows that the stress superimposition effect is larger at the central area. The earth pressure under the buildings fluctuates by the disturbance of the pile tip. The results from the 3D numerical simulation match well with the centrifugal model test which can be used to analyze and to predict the land subsidence of high-rise buildings.

(4) In the general model test, this monograph studied the deformation law of each soil layer considering dual effects of the building load and the groundwater withdrawal. The later constructed building accelerates the subsidence rate of soil layers having been under the load of the earlier constructed building, while the earlier constructed building reduces the subsidence of soil layers under the later constructed building to some extent, for the soil foundation under the later constructed building has already consolidated under the earlier constructed building load for some time. For the superimposition effect of land subsidence caused by buildings, the subsidence inside the building area increases while the differential subsidence outside the building area increases. The dewatering changes the subsidence distribution which develops at the building load period. When dewatering, the friction force from piles reduces the subsidence of buildings and the surrounding area, yet increases the differential subsidence. Compared with the subsidence caused by the building loads, the subsidence caused by the dewatering develops slowly and lasts for a longer period of time. The compression of soft soil layers accounts for the most part of land subsidence at both the building load period and the dewatering periods. Groundwater recharge can mitigate land subsidence effectively, yet the rebound is little. It is not appropriate to estimate the land subsidence by the net groundwater pumping. If the net amount of groundwater pumping is zero at a certain time, it cannot be determined that land subsidence has no relationship with the groundwater pumping. The subsidence may be caused by the early pumping.

(5) The engineering characteristics of soils are controlled by the state of the microstructure of soils to a great extent. Flocculation and honeycomb-flocculation are the main structures in silty clay of No. 4 and clayey soil of layer No. 8 and most of the structures are laminar, which gives birth to the overhead structure with high porosity. The particles are rearranged and consolidation is conducted by the building loads and the stress superimposition effect. The microstructure of clay determines the mechanics characteristics for engineering and macro-land subsidence of layer No. 4 and layer No. 8 is larger. After the test, the total porosity decreases. The mineral of sand is mainly composed of quartz, including some feldspar and mica, etc. The hardness of quartz is large. So the sand layer is usually used as the bearing stratum of the pile tip for the high-rise building in the soft area. Because of better mechanics characteristics for engineering, the subsidence of layer No. 7 is smaller. The microstructure of silty sand is destroyed a little by the building load and the stress superimposition effect. After the test, the total porosity decreases. There are mainly macropores in the layer No. 4 silty clay, the layer No. 8 clayey soil, the layer No. 7 silty sand, and the layer No. 9 silty sand in Shanghai. The ink-bottle effect exists in the intrusion stage in the MIP test. The pore size distribution of each soil layer was studied by the fractal theory. There are four different fractal dimensions in layer No. 4 silty clay and layer No. 8 clayey soil. There are three different fractal dimensions in layer No. 7 silty sand and layer No. 9 silty sand in Shanghai. The parameter of specific subsidence is put forward as a tie to analyze the relationship between land subsidence and pore

structure of soils. For the silt sand layer (layer No. 7 and layer No. 9), the mean pore size, the total specific surface area, the total porosity, and the holding mercury coefficient decrease with the specific subsidence increasing. For the clayey soil (layer No. 4 and layer No. 8), the mean pore size, the total porosity, and the holding mercury coefficient increase with the specific subsidence increasing, but the total pore specific surface area decreases with the specific subsidence increasing.

(6) The adopted threshold determines the calculated void ratio from the SEM images. The larger the adopted threshold, the smaller the void ratio will be. The real filling rate n' can be estimated by adding about 15% to the filling rate calculated with the Otsu method. The calculated filling rate and the adopted threshold T_n are in good agreement with the Gaussian function relationship. When calculating the void ratio with SEM images, the appropriate amplifications of images representing the soil microstructure should be chosen. 3000× or 5000× for the fine grain soil and 500× or 1000× for the coarse grain soil. The selected T_n should be between the range [0.34, 0.36], which corresponds to the range [85, 90] of the real grayscale value. After the consolidation, the total pore area of the silty clay of layer No. 4 and the clayey soil of layer No. 8 decreased greatly, the amplitudes of reduction are 28.0 and 29.6%, respectively, showing the great compressibility of the two soil layers and are coincident with the large macroscopic subsidence. The pore area of the pores whose radii are bigger than 1 μm occupies at least 97% of the total pore area. Before the consolidation, the orienting probability entropy of the silty clay of layer No. 4 and the clayey soil of layer No. 8 was big, meaning the large degree of the disorder of the pore structure. After the consolidation, the degree was even bigger, showing the soil consolidation process, that is, the soils were compressed and the pores become smaller. Some longer pores were even compressed to produce more pores, which induces the dispersal of the orientation of the long axis. After the consolidation, the fractal dimension of shape D of the pores in the silty clay of layer No. 4 and the clayey soil of layer No. 8 decreased. Considering the changes of the average eccentricity and the average shape factor, it can be seen that during the soil consolidation, the pores are not simply squashed but parts of the boundaries are compressed to be closed, which makes the shape of pores more regular.

(7) The concept of the geology-environmental capacity of the buildings is put forward in this paper. By analyzing the main factors of land subsidence and building up the evaluation system, the geology-environmental capacity of the ground building comes to be quantitative. The geological environmental capacity of the ground building is mainly controlled by the land subsidence and the output is the floor area ratio. According to the different geology structures and the different requirements of subsidence control in the soft soil area in Shanghai, the evaluation system of the floor area ratio is built up by the adaptive neuro-fuzzy inference system (ANFIS) and the floor area ratios of four typical regions are obtained by the ANFIS to offer a reference to the urban planning. If the floor area ratios of the regions are bigger, the high-rise

buildings are dense and land subsidence of the regions is larger. It is adverse to continue to plan the high-rise construction in these regions. However, the floor area ratios of the regions are smaller and land subsidence is less and the high-rise buildings can be planned for construction.

10.2 Prospects for Further Study

This monograph studied the land subsidence caused by the engineering-environmental effect including the loads of high-rise building and the changing of groundwater table. But there is still some work to be further studied.

(1) The engineering geology of the site for groundwater exploitation is complex and the horizontal dimension is much larger than the longitudinal dimension. The conventional finite-element method for grid computing is very large. The multi-scale finite-element method can be used to analyze the land subsidence at the scene.
(2) The stress level of the model test is low, and the stress state of the soil in situ cannot be reduced. Centrifugal model test can be taken into consideration when the steady duration of pumping consolidation is long and the conditions are allowed.

Appendix A
The Element of Matrix Φ

$$
\begin{aligned}
\Phi_{11} = {} & (b_{21}b_{33} + a_{13}a_{25}b_{31} + a_{25}b_{23}b_{31} - 4a^2a_{51}b_{33} - a^4b_{21}b_{33} - 2a^2a_{13}a_{25}b_{31} \\
& + a^4a_{13}a_{25}b_{31} - a^4a_{25}b_{23}b_{31}) \Big/ \left[a_{25}b_{33}(a^2-1)^2(b_{11}+b_{21})\right] \\
& + (2a^4a_{13}a_{25}b_{31} - 4a^2a_{13}a_{25}b_{31} \\
& + 2a_{13}a_{25}b_{31}) \Big/ \left[a_{25}b_{33}(a^2-1)^2(b^2-1)(b_{11}+b_{21})\right]; \\
\Phi_{12} = {} & (2a^4a_{13}a_{25}b_{31} - 4a^2a_{13}a_{25}b_{31} \\
& + 2a_{13}a_{25}b_{31}) \Big/ \left[b_{33}(a^2-1)^2(b^2-1)(b_{11}+b_{21})\right] \\
& - (b_{11}b_{33} - a_{13}a_{25}b_{31} - a_{25}b_{23}b_{31} + 4a^2a_{51}b_{33} - a^4b_{11}b_{33} + 2a^2a_{13}a_{25}b_{31} \\
& - a^4a_{13}a_{25}b_{31} + a^4a_{25}b_{23}b_{31}) \Big/ \left[b_{33}(a^2-1)^2(b_{11}+b_{21})\right]; \\
\Phi_{13} = {} & (a_{13} + b_{23} + a_{13}b^2 - b^2b_{23}) / \left[b_{33}(b^2-1)\right] \\
& + (-2b_{23}b^2 + 2b_{23}) / \left[b_{33}(a^2-1)(b^2-1)\right]; \\
\Phi_{14} = {} & (2aa_{51}b_{33} - 2ab_{21}b_{33} + 2a^3a_{51}b_{33} + 2a^3b_{21}b_{33} + 2a^3a_{25}b_{23}b_{31} \\
& - 2aa_{25}b_{23}b_{31}) \Big/ \left[a_{25}b_{33}(a^2-1)^2(b_{11}+b_{21})\right] \\
& - (b(2a^4a_{13}a_{25}b_{31} - 4a^2a_{13}a_{25}b_{31} \\
& + 2a_{13}a_{25}b_{31})) \Big/ \left[a_{25}b_{33}(a^2-1)^2(b^2-1)(b_{11}+b_{21})\right]; \\
\Phi_{15} = {} & (2aa_{51}b_{33} + 2ab_{11}b_{33} + 2a^3a_{51}b_{33} - 2a^3b_{11}b_{33} \\
& + 2a^3a_{25}b_{23}b_{31} - 2aa_{25}b_{23}b_{31}) \Big/ \left[b_{33}(a^2-1)^2(b_{11}+b_{21})\right] \\
& - (b(2a^4a_{13}a_{25}b_{31} - 4a^2a_{13}a_{25}b_{31} \\
& + 2a_{13}a_{25}b_{31})) \Big/ \left[b_{33}(a^2-1)^2(b^2-1)(b_{11}+b_{21})\right]; \\
\Phi_{16} = {} & 2a_{13}b/(-b_{33}b^2 + b_{33}) - 2ab_{23}/(-b_{33}a^2 + b_{33});
\end{aligned}
$$

Z.-D. Cui, *Land Subsidence Induced by the Engineering-Environmental Effect*,
https://doi.org/10.1007/978-981-10-8040-1

$$\Phi_{21} = \left(a_{21}b_{33} - b_{21}b_{33} - a_{13}a_{25}b_{31} - a_{25}b_{23}b_{31} - a^4a_{21}b_{33} + 4a^2a_{51}b_{33}\right.$$
$$+ a^4b_{21}b_{33} + 2a^2a_{13}a_{25}b_{31}$$
$$\left.-a^4a_{13}a_{25}b_{31} + a^4a_{25}b_{23}b_{31}\right)\Big/\left[b_{33}(a^2-1)^2(b_{11}+b_{21})\right]$$
$$-(2a^4a_{13}a_{25}b_{31} - 4a^2a_{13}a_{25}b_{31} + 2a_{13}a_{25}b_{31})\Big/\left[b_{33}(a^2-1)^2(b^2-1)(b_{11}+b_{21})\right];$$

$$\Phi_{22} = -\left[a_{25}(a_{21}b_{33} + b_{11}b_{33} + a_{13}a_{25}b_{31} - a_{25}b_{23}b_{31}\right.$$
$$- a^4a_{21}b_{33} + 4a^2a_{51}b_{33} - a^4b_{11}b_{33} - a_{21}b^2b_{33} - b^2b_{11}b_{33}$$
$$- 2a^2a_{13}a_{25}b_{31} + a^4a_{13}a_{25}b_{31} + a_{13}a_{25}b^2b_{31}$$
$$+ a^4a_{25}b_{23}b_{31} + a_{25}b^2b_{23}b_{31} + a^4a_{21}b^2b_{33} - 4a^2a_{51}b^2b_{33}$$
$$+ a^4b^2b_{11}b_{33} - 2a^2a_{13}a_{25}b^2b_{31} + a^4a_{13}a_{25}b^2b_{31}$$
$$\left.-a^4a_{25}b^2b_{23}b_{31})\right]\Big/\left[b_{33}(a^2-1)^2(b^2-1)(b_{11}+b_{21})\right];$$

$$\Phi_{23} = a_{13}a_{25}/[b_{33}(b+1)] - a_{13}a_{25}/[b_{33}(b-1)]$$
$$+ a_{25}(a_{13} + b_{23} - a^2a_{13} + a^2b_{23})/[b_{33}(b+1)];$$

$$\Phi_{25} = 2a_{25}b(a^4a_{13}a_{25}b_{31}$$
$$- 2a^2a_{13}a_{25}b_{31} + a_{13}a_{25}b_{31})\Big/\left[b_{33}(a^2-1)^2(b^2-1)(b_{11}+b_{21})\right]$$
$$- 2a_{25}(aa_{21}b_{33} + aa_{51}b_{33} + ab_{11}b_{33} - a^3a_{21}b_{33} + a^3a_{51}b_{33} - a^3b_{11}b_{33}$$
$$+ a^3a_{25}b_{23}b_{31} - aa_{25}b_{23}b_{31})\Big/\left[b_{33}(a^2-1)^2(b_{11}+b_{21})\right];$$

$$\Phi_{26} = 2aa_{25}b_{23}/(-b_{33}a^2 + b_{33}) - 2a_{13}a_{25}b/(-b_{33}b^2 + b_{33});$$

$$\Phi_{31} = \left(b_{31} - b_{33} - a^2b_{31} - a^2b_{33} + b^2b_{31} + b^2b_{33}\right.$$
$$\left.-a^2b^2b_{31} + a^2b^2b_{33}\right)/\left[b_{33}(a^2-1)(b^2-1)(b_{11}+b_{21})\right];$$

$$\Phi_{32} = -a_{25}(-a^2b^2 + a^2 - b^2 + 1)/\left[(a^2-1)(b^2-1)(b_{11}+b_{21})\right]$$
$$-a_{25}(b_{31}b^2 + b_{31})/\left[b_{33}(b^2-1)(b_{11}+b_{21})\right];$$

$$\Phi_{33} = -(b^2+1)/\left[b_{33}(b^2-1)\right];$$

$$\Phi_{34} = 2(b_{31}a^2b - b_{33}ab^2 + b_{33}a - b_{31}b)/\left[b_{33}(a^2-1)(b^2-1)(b_{11}+b_{21})\right];$$

$$\Phi_{35} = 2a_{25}(a^2b_{31}b - b_{33}ab^2 + b_{33}a - b_{31}b)/\left[b_{33}(a^2-1)(b^2-1)(b_{11}+b_{21})\right];$$

$$\Phi_{36} = 2b/\left[b_{33}(b^2-1)\right];$$

$$\Phi_{42} = \left(2aa_{51}b_{33} + 2ab_{11}b_{33} + 2a^3a_{51}b33 - 2a^3b_{11}b_{33} + 2a^3a_{25}b_{23}b_{31} - 2aa_{25}b_{23}b_{31}\right)\Big/\left[b_{33}(a^2-1)^2(b_{11}+b_{21})\right] - b(2a_{13}a_{25}b_{31}a^4 - 4a^2a_{13}a_{25}b_{31} + 2a_{13}a_{25}b_{31})\Big/\left[b_{33}(a^2-1)^2(b^2-1)(b_{11}+b_{21})\right];$$

$\Phi_{24} = -\Phi_{15};$

$\Phi_{41} = \Phi_{14};$

$\Phi_{43} = \Phi_{16};$

$\Phi_{44} = \Phi_{11};$

$\Phi_{45} = -\Phi_{12};$

$\Phi_{46} = \Phi_{13};$

$\Phi_{51} = -\Phi_{42};$

$\Phi_{52} = \Phi_{25};$

$\Phi_{53} = \Phi_{26};$

$\Phi_{54} = \Phi_{21};$

$\Phi_{55} = \Phi_{22};$

$\Phi_{56} = \Phi_{23};$

$\Phi_{61} = \Phi_{34};$

$\Phi_{62} = \Phi_{35};$

$\Phi_{63} = \Phi_{36};$

$\Phi_{64} = \Phi_{31};$

$\Phi_{65} = \Phi_{32};$

$\Phi_{66} = \Phi_{33};$

where $a = e^{-\xi z}$; $b = e^{-qz}$; $a_{13} = \frac{\xi}{M\left(-q^2+\xi^2\right)}$; $a_{21} = 1 - M\varphi$; $a_{25} = 2G\xi$; $a_{41} = \frac{1-(M-G)\varphi}{2G}z$; $a_{51} = z\xi(1 + G\varphi - M\varphi)$; $b_{11} = G\varphi$; $b_{21} = 1 - G\varphi - M\varphi$; $b_{23} = \frac{q}{M\left(q^2-\xi^2\right)}$; $b_{31} = \xi k_z$; $b_{33} = qk_z$.

Appendix B
The Adaptive Neuro-Fuzzy Inference System (ANFIS)

If x is A and y is B, then $Z = f(x, y)$, where A and B are the fuzzy set of the inputs and $z = f(x, y)$ is the accurate function of the target. Generally, $f(x, y)$ is polynomial about input variables x, y. If $f(x, y)$ is one-order polynomial, the obtained fuzzy inference system is the one-order Sugeno fuzzy model.

Figure B.1 shows the inference process of the one-order Sugeno fuzzy model which has two inputs x, y and one output z, such that it contains two fuzzy "if-then" rules:

If x is A_1 and y is B_1, then $f_1 = p_1x + q_1y + r_1$

If x is A_1 and y is B_2, then $f_2 = p_2x + q_2y + r_2$

where A_i and B_i are fuzzy sets corresponding with input variables.

Supposing that the S type membership functions of input variables x and y are:

$$S_{A_i}(x, a_i, b_i) = \frac{1}{1 + e^{-a_i(x-b_i)}}$$

$$S_{B_i}(y, c_i, d_i) = \frac{1}{1 + e^{-c_i(x-d_i)}}$$

where i = 1,2, $\{a_i, b_i\}$ and $\{c_i, d_i\}$ being two group characteristic parameters of S type membership functions.

S type membership functions are changed with the change of the values of characteristic parameters, that is, the membership functions of A_i and B_i are changed. The inference process can be equivalent to the ANFIS structure shown in Fig. B.2 which consists of five layers.

The function of the first layer is to compute the fuzzy membership of inputs. Every node of this layer is the adaptive node and has a node function,

$$O_{1,i} = S_{A_i}(x, a_i, b_i), \quad i = 1,\ 2$$

$$O_{1,j} = S_{B_{j-2}}(y, c_{j-2}, d_{j-2}), \quad j = 3,\ 4$$

Z.-D. Cui, *Land Subsidence Induced by the Engineering-Environmental Effect*,
https://doi.org/10.1007/978-981-10-8040-1

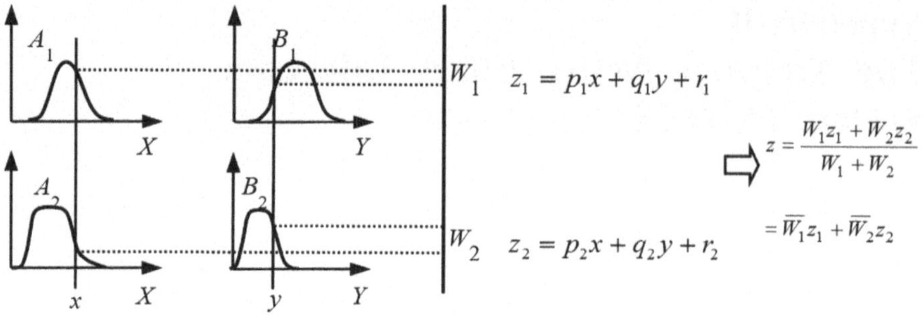

Figure B.1 The inference process of one-order Sugeno fuzzy system

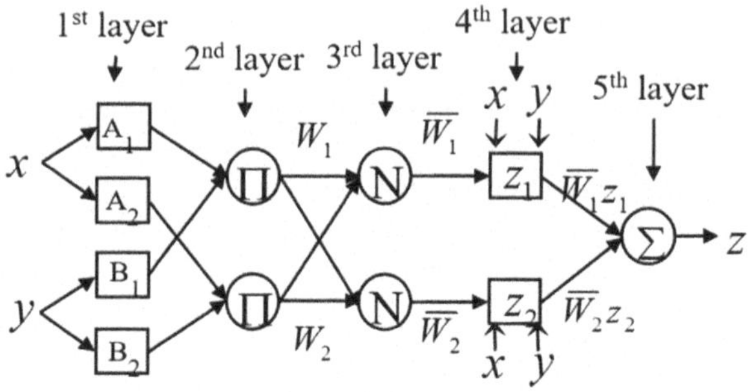

Figure B.2 The ANFIS structure equivalent to one-order Sugeno

where $O_{1,i}$ is the No. i output of the first layer and is a member of the corresponding output variable of fuzzy sets A_i and B_i.

The function of the second layer is to compute the fitness of every rule. Every node of this layer is a fixed node labeled Π. Its output is the product of all input signals:

$$O_{2,1} = O_{1,1} \cdot O_{1,3} = S_{A_1}(x, a_1, b_1) \cdot S_{B_1}(y, c_1, d_1), \quad \text{mariking } W_1$$
$$O_{2,2} = O_{1,2} \cdot O_{1,4} = S_{A_2}(x, a_2, b_2) \cdot S_{B_2}(y, c_2, d_2), \quad \text{marking } W_2$$

The function of the third layer is to compute the unitary value of fitness. Every node of this layer is a fixed node labeled N. The ratio of the strength of one rule to the sum of all the strengths is obtained:

$$O_{3,1} = \overline{W}_1 = \frac{W_1}{W_1 + W_2}$$
$$O_{3,2} = \overline{W}_2 = \frac{W_2}{W_1 + W_2}$$

The function of the fourth layer is to compute the output of every rule. Every node of this layer is an adaptive node with a function:

$$O_{4,1} = \overline{W}_1 z_1 = \overline{W}_1 (p_1 x + q_1 y + r_1)$$
$$O_{4,2} = \overline{W}_2 z_2 = \overline{W}_2 (p_2 x + q_2 y + r_2)$$

where $\{p_i, q_i, r_i\}$ $(i = 1, 2)$ is the set of parameters of corresponding nodes, ie conclusion parameters.

The function of the fifth layer is to compute the output of the fuzzy system. The single node of this layer is a fixed node (Σ) which compute the sum of all the incoming signals:

$$O_5 = z = \sum \overline{W}_i z_i = \overline{W}_1 z_1 + \overline{W}_2 z_2$$

This network consists of undetermined characteristic parameters (a_i, b_i, c_i, $d_i(i = 1, 2)$ of membership functions and p_i, q_i, $r_i(i = 1, 2)$ of conclusion parameters). ANFIS dynamically adjusts these parameters in the trained process.

Appendix C
Major Published Works of the Book Author

English Journal Papers

1. **Cui ZD***, Jia YJ (2018) Physical model test of layered soil subsidence considering dual effects of building load and groundwater withdrawal. Arab J Sci Eng (SCI). https://doi.org/10.1007/s13369-017-2716-7
2. **Cui ZD***, Yuan Q, Yang JQ (2018) Laboratory model tests about the sand embankment supported by piles with a cap beam. Geomech Geoengineering Int J 13(1):64–76
3. Zhang ZL, **Cui ZD*** (2018) Effect of freezing-thawing on dynamic characteristics of the silty clay under K0-consolidated condition. Cold Reg Sci Technol (SCI) 146:32–42
4. Zhang ZL, **Cui ZD*** (2018) Effects of freezing-thawing and cyclic loading on pore size distribution of silty clay by mercury intrusion porosimetry. Cold Reg Sci Technol 145:185–196
5. Yuan L, **Cui ZD***, Tan J (2017) Numerical simulation of longitudinal settlement of shield tunnel in the coastal city, Shanghai. Mar Georesour Geotechnol (SCI) 35(3):365–370
6. Li Z, **Cui ZD*** (2017) Axisymmetric consolidation of saturated multi-layered soils with anisotropic permeability due to well pumping. Comput Geotech (SCI) 92:229–239
7. Zhang CL, **Cui ZD*** (2017) Numerical simulation of dynamic response around shield tunnel in the soft soil area. Mar Georesour Geotechnol (SCI) 35 (7):1018–1027
8. Zhang ZL, **Cui ZD*** (2017) Analysis of microscopic pore structures of the silty clay before and after freezing–thawing under the subway vibration loading. Environ Earth Sci (SCI) 76:528
9. **Cui ZD***, Jia YJ, Yuan L (2016) Distribution law of soil deformation caused by decompression of confined water. Environ Earth Sci (SCI) 75(18):1281
10. **Cui ZD***, Li Z, Jia YJ (2016) Model test study on the subsidence of high-rise building group due to the variation of groundwater level. Nat Hazards (SCI) 84 (1):35–53

Z.-D. Cui, *Land Subsidence Induced by the Engineering-Environmental Effect*,
https://doi.org/10.1007/978-981-10-8040-1

11. **Cui ZD***, Zhao LZ, Yuan L (2016) Microstructures of consolidated Kaolin clay at different depths in centrifuge model tests. Carbonates Evaporites (SCI) 31 (1):47–60
12. Tan J, **Cui ZD***, Yuan Y (2016) Study on the long-term settlement of subway tunnel in soft soil area. Mar Georesour Geotechnol (SCI) 34(5):486–492
13. Yuan Q, **Cui ZD*** (2016) Two-dimensional numerical analysis of the subgrade improved by stone columns in the soft soil area. Mar Georesour Geotechnol (SCI) 34(1):79–86
14. **Cui ZD***, Yang JQ, Yuan L (2015) Land subsidence caused by the interaction of high-rise buildings in soft soil areas. Nat Hazards (SCI) 79(2):1199–1217
15. **Cui ZD***, Zhang ZL (2015) Comparison of dynamic characteristics of the silty clay before and after freezing and thawing under the subway vibration loading. Cold Reg Sci Technol (SCI) 119:29–36
16. **Cui ZD***, Yuan Q (2015) Investigation of the subsidence induced by the Maglev train. Nat Hazards (SCI) 75(2):1767–1778
17. He PP, **Cui ZD*** (2015) Dynamic response of a thawing soil around the tunnel under the vibration load of subway. Environ Earth Sci (SCI) 73(5):2473–2482
18. **Cui ZD***, Tan J (2015) Study on settlements of the running tunnels and subway stations based on the in-situ monitoring data. Nat Hazards (SCI) 75(1):465–472
19. **Cui ZD***, Ren SX (2014) Prediction of long-term settlements of subway tunnel in the soft soil area. Nat Hazards (SCI) 74(2):1007–1020
20. **Cui ZD***, He PP, Yang WH (2014) Mechanical properties of a silty clay subjected to freezing-thawing. Cold Reg Sci Technol (SCI) 98:26–34
21. **Cui ZD***, Jia YJ (2013) Analysis of electron microscope images of soil pore structure for the study of land subsidence in centrifuge model tests of high-rise building groups. Engineering Geology (SCI), 164: 107-116.
22. Song L, **Cui ZD**, Zhang HQ (2013) Analysis and treatment of the fault activation below the dynamic foundation in the goaf area. Disaster Adv (SCI) 6 (S1):337–342
23. Yuan L, **Cui ZD** (2013) Reliability analysis for the consecutive-k-out-of-n: F system with repairmen taking multiple vacations. Appl Math Model (SCI), 37 (7):4685–4697
24. **Cui ZD***, Jia YJ (2012) Study on the mechanisms of the soil consolidation and land subsidence caused by the high-rise building group in the soft soil area. Disaster Adv (SCI) 5(4):604–608
25. **Cui ZD***, Ren SX (2012) Long-term deformation characteristics of the soil around the subway tunnel induced by the vibration loading. Disaster Adv (SCI) 5(4):1791–1797
26. **Cui ZD*** (2012) Bearing capacity of single pile and in-flight T-Bar Penetration for centrifuge modeling of land subsidence caused by the interaction of high-rise buildings. Bull Eng Geol Environ (SCI) 71(3):579–586
27. **Cui ZD*** (2012) Land subsidence disaster caused by natural factors and human activities. Disaster Adv (SCI) 5(2):3–4
28. **Cui ZD*** (2011) Effect of water–silt composite blasting on the stability of rocks surrounding a tunnel. Bull Eng Geol Environ (SCI) 70(4):657–664

29. **Cui ZD***, Tang YQ (2011) Microstructures of different soil layers caused by the high-rise building group in Shanghai. Environ Earth Sci (SCI) 63(1):109–119
30. **Cui ZD***, Wang HM (2010) Land subsidence at different points among a group of high-rise buildings. Disaster Adv (SCI) 3(4):63–66
31. **Cui ZD***, Yuan L, Yan CL (2010) Water-silt composite blasting for tunneling. Int J Rock Mech Min Sci (SCI) 47(6):1034–1037
32. **Cui ZD***, Tang YQ (2010) Land subsidence and pore structure of soils caused by the high-rise building group through centrifuge model test. Eng Geol (SCI) 113(1–4):44–52
33. **Cui ZD***, Tang YQ, Yan XX (2010) Centrifuge modeling of land subsidence caused by the high-rise building group in the soft soil area. Environ Earth Sci (SCI) 59(8):1819–1826
34. **Cui ZD***, Tang YQ, Yan XX, Wang HM, Yan CL, Wang JX (2010) Evaluation of the geology-environmental capacity of buildings based on the ANFIS model of the floor area ratio. Bull Eng Geol Environ (SCI) 69(1):111–118
35. **Cui ZD***, Tang YQ, Guo CQ, Yuan L, Yan CL (2008) Flow-induced vibration and stability of an element model for parallel-plate fuel assemblies. Nuclear Eng Des (SCI) 238(7):1629–1635
36. Tang YQ, **Cui ZD**, Wang JX, Lu C, Yan XX (2008) Model test study of land subsidence caused by high-rise building group. Bull Eng Geol Environ (SCI) 67(2):173–179
37. Tang YQ, **Cui ZD**, Wang JX, Yan LP, Yan XX (2008) Application of grey theory-based model to prediction of land subsidence due to engineering environment in Shanghai. Environ Geol (SCI) 55(3):583–593
38. Tang YQ, **Cui ZD**, Zhang X, Zhao SK (2008) Dynamic response and pore-water pressure model of saturated soft clay around a tunnel induced by the subway vibration load. Eng Geol (SCI) 98(3–4):126–132

English Conference Papers

1. **Cui ZD***, Hua SS (2018) Long-term settlement of subway tunnel and prediction of settlement trough in coastal city Shanghai. Geoshanghai, Shanghai
2. **Cui ZD***, Fan SC (2018) Physical modeling of arching effect in the piled embankment. Geoshanghai, Shanghai
3. Zhang ZL, **Cui ZD*** (2018) Dynamic response of soil around the tunnel under subway vibration loading. Geoshanghai, Shanghai
4. **Cui ZD***, Zhang ZL (2017) Comparison of the dynamic characteristics of the silty clay before and after freezing-thawing under cyclic loadings. 19th ICSMGE, Seoul, pp 335–339

5. **Cui ZD***, Zhang CL, Hou CY (2017) Vibrations induced by subway moving load with excitation of the track vertical profile irregularity. 15th IACMAG, Wuhan pp 1–18
6. **Cui ZD***, Yuan L, Yan LP, Guo CQ (2009) Stability analysis and measurement of flow-induced vibration of a parallel-plate structure. International Conference on Earthquake Engineering-the 1st Anniversary of Wenchuan Earthquake (ISTP), Chengdu pp 288–293
7. **Cui ZD**, Tang YQ, Zhang X (2008) Deformation and pore pressure model of saturated soft clay around a subway tunnel. The Six International Symposium Geotechnical Aspects of Underground Construction in Soft Ground (EI), Shanghai pp 769–774
8. Tang YQ, **Cui ZD**, Zhang X (2008) Dynamic response of saturated silty clay around a tunnel under subway vibration loading in Shanghai. The Six International Symposium Geotechnical Aspects of Underground Construction in Soft Ground (EI), Shanghai pp 843–848

*Noted the corresponding author.

The manufacturer's authorised representative in the EU is Springer Nature Customer Service Centre GmbH, Europaplatz 3, 69115 Heidelberg, Germany. If you have any concerns regarding our products, please contact ProductSafety@springernature.com

Printed and bound by CPI Group (UK) Ltd, Croydon, CR0 4YY
30/09/2026
02225448-0006